# 5th International Field Symposium "West Siberian Peatlands and Carbon Cycle: Past and Present" and the International Conference "Carbon Balance of Western Siberian Mires in the Context of Climate Change"

IOP Conference Series: Earth and Environmental Science
Volume 138



Khanty-Mansiysk, Russia
19 – 29 June 2017

ISBN: 978-1-5108-6061-2
ISSN: 1755-1307

**Printed from e-media with permission by:**

Curran Associates, Inc.
57 Morehouse Lane
Red Hook, NY 12571

**Some format issues inherent in the e-media version may also appear in this print version.**



Printed by Curran Associates, Inc. (2018)

For permission requests, please contact the Institute of Physics
at the address below.

Institute of Physics
Dirac House, Temple Back
Bristol BS1 6BE UK

Phone: 44 1 17 929 7481
Fax: 44 1 17 920 0979

techtracking@iop.org

**Additional copies of this publication are available from:**

Curran Associates, Inc.
57 Morehouse Lane
Red Hook, NY 12571 USA
Phone: 845-758-0400
Fax: 845-758-2633
Email: curran@proceedings.com
Web: www.proceedings.com

# Table of contents

## Volume 138

**5th International field symposium "West Siberian Peatlands and Carbon Cycle: Past and Present" and the International conference "Carbon Balance of Western Siberian Mires in the Context of Climate Change"**

**19–29 June 2017, Khanty-Mansiysk, Russia**

**Accepted papers received: 15 March 2018**
**Published online: 28 March 2018**

### Preface



### Papers

IOP Conf. Series: Earth and Environmental Science **138** (2018) 011001 doi:10.1088/1755-1315/138/1/011001

# Preliminary material

## 1. Preface

The volume contains papers presented at the Fifth International Field Symposium «West Siberian Peatlands and Carbon Cycle: Past and Present» and the International conference «Carbon Balance of Western Siberian Mires in the Context of Climate Change» 19-29 June, 2017 Khanty-Mansiysk, Russia. Published papers cover diverse areas of mire study and allied disciplines. The issues of biodiversity of mire flora and fauna, problems of the mire genesis, evolution and paleoecology are considered. The carbon balance in mire ecosystems is evaluated in relation to new approaches and data, which allows understanding the nature of current climate change and making decisions basing on reliable scientific knowledge.

Materials of the symposium and the conference are of interest for researchers of mire study, biogeochemistry, ecology, soil science, environmental protection, nature conservation.

## 2. Organizing committee

### 2.1. *Chairman of the organizing committee*

Andrey Sirin, director of the Institute of Forest Science, Russian Academy of Sciences.

### 2.2. *Organizers*

Yugra State University, Khanty-Mansiysk
University of Orleans, France
Institute of Forest Science, Russian Academy of Sciences, Moscow
Institute of Soil Science and Agrochemistry of the Siberian Branch of the RAS, Novosibirsk
Arctic Research Center of the Yamal-Nenets autonomous district, Salekhard
National Research Tomsk State University

## 3. Organizational and financial support

The Government of Khanty-Mansiysk Autonomous Okrug – Ugra
National Center for Scientific Research (CNRS), France
International project «Biogeochemical cycles of Arctic wetlands of Western Siberia as an indicator of climate change on a global scale and the basis for environmental management in the region (BIO-GEOCLIM)»
International project «International Network for Terrestrial Research and Monitoring in the Arctic, INTERACT»
Russian Foundation for Basic Research (Grants 17-05-20260, 16-55-16007)
LabInstruments Ltd

# Peer review statement

All papers published in this volume of *IOP Conference Series: Earth and Environmental Science* have been peer reviewed through processes administered by the proceedings Editors. Reviews were conducted by expert referees to the professional and scientific standards expected of a proceedings journal published by IOP Publishing.

IOP Conf. Series: Earth and Environmental Science **138** (2018) 012001 doi:10.1088/1755-1315/138/1/012001

# Reconstruction of postglacial landscape evolution within the eastern periphery of Chuya depression on the basis of multidisciplinary analysis of peats in Boguty river basin, SE Altai, Russia

**A R Agatova*[1,2], I V Khazina[3], M A Bronnikova[4], O N Uspenskaya[5], R K Nepop[1,2]**

[1] Sobolev Institute of Geology and Mineralogy SB RAS, Koptyuga av., 3, Novosibirsk, Russia
[2] Ural Federal University, Mira str., 19, Yekaterinburg, Russia
[3] Trofimuk Institute of Petroleum Geology and Geophysics SB RAS, Koptyuga av., 3, Novosibirsk, Russia
[4] Institute of Geography RAS, Staromonetnyj, 29, Moscow, Russia
[5] Russian Research Institute of Vegetable Growing, Vereya-500, Moscow

*Email: agatr@mail.ru

**Abstract**. This paper presents the results of multidisciplinary investigations of the peat-bed under hummocky permafrosted boggy meadow within the Boguty basin. For the last 7600 years 4 evolutionary phases of peatbog formation within the drained part of Low Boguty Lake bottom were established and corresponding 4 pollen complexes were described. 18 radiocarbon dates suggest some chronological bench marks of postglacial landscape evolution in the region. After degradation of Sartan glaciation about 14000 BP, trees grew in now forestless areas at 11000 BP and 8500-7800 BP. The climate in the first half of the Holocene was warmer and more humid. Accumulation of lacustrine loams within the studied peatbog occurred before 7600 BP with predominated algae Pediastrum, Zygnemataceae and Botryococcus. After the lake level lowered in the result of destructing moraine dam, two lower peat horizons were developed about 7600 – 7200 BP. An episode of significant lake desiccation (later than 7200 BP) was recorded in all proxy archives. Further rise of water supply led to increasing the number of water-bog plants, diatomaceous, euglenic and green algae. At the same time, the pollen of xerophytes began to predominate in the pollen complexes, indicating aridization. The final stage reflects stable peatbog drying and its transformation into boggy meadow, decomposition and mineralization of peat.

## 1. Introduction

In the framework of our multidisciplinary investigations of paleoenvironmental changes in the southeastern part of the Russian Altai (SE Altai), and in order to reconstruct the postglacial climatically driven landscape evolution within the eastern periphery of the Chuya intermountain depression, the hummocky permafrosted boggy meadow was studied in the high mountainous Boguty basin. This area bordering with the Mongolian Altai is significantly less investigated in comparison with central parts of the Chuya depression – the largest intermountain basin in the SE Altai. In addition to the first geological and geomorphological studies [1, 2], archaeological sites of Paleolithic time, Late Bronze Age and Iron Epoch are described here (brief review is given in [3]). It makes multidisciplinary regional investigations, including presented in this paper analysis of peatbog, relevant also for reconstructing the Holocene chronology and paleoenvironmental context of nomadic societies settling in this high-mountain area.

## 2. Study area

The Boguty river belongs to the basin of the Chuya river - the main water artery of the SE Altai. The Boguty river basin (about 35 km long) includes a part of western slope of the Chikhachev range and small (about 25x13 km) Boguty depression, which forms an eastern periphery of the Chuya intermountain depression (figure 1). The Chikhachev range stretches in south-north direction at about 100 km, and has maximal altitudes up to 4029 m a.s.l. in its southern part. Within the Boguty basin maximal altitudes are about 3550-3700 m a.s.l. Beginning in the axial part of the ridge, the Right and Left Boguty at their confluence form the Boguty river. Downstream, within the Boguty depression, the Boguty river connected three largest lakes in its basin – Upper Boguty (or Boguty according some maps), Middle Boguty (or Aday) and Low Boguty (or Kok-Kul).

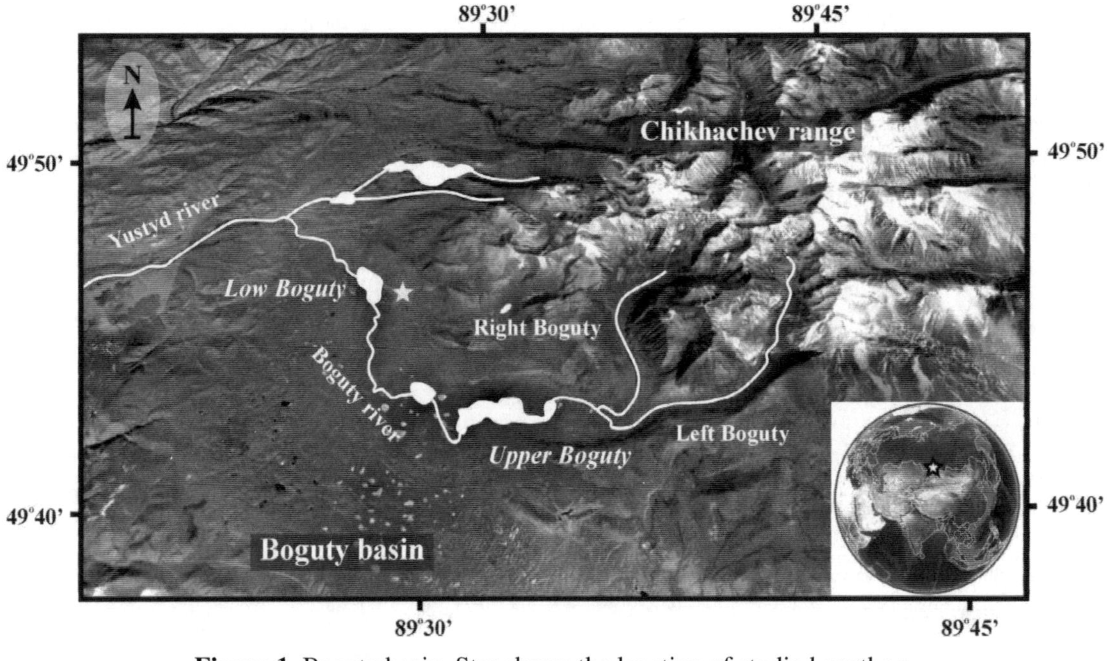


**Figure 1.** Boguty basin. Star shows the location of studied peatbog.

The area is characterized by extreme ultra continental cryoarid permafrost affected environment. Within the Boguty basin mountain steppe is changing with height into tundra-steppe and tundra vegetation. The forests, which have an insular distribution in the western mountains flanking the Chuya depression, are completely vanishing in its eastern periphery, and there is no forest vegetation at all within the studied area. The higher topographic belt is represented by alpine landscapes and glacial zone. The modern glaciation within the axial parts of the ridges is practically completely degraded [4].

In the Late Pleistocene the western slope of the Chikhachev range and adjusted Boguty depression were occupied by ice-sheet. Its degradation was accompanied by formation of terminal moraine complexes and moraine-dammed lakes. Further destruction of dams was followed by water level lowering and fluvial system transformation, as well as by developing thermokarst lakes and peatbogs within drying bottom of the basin and by spreading vegetation in it.

The studied peatbog is located in the eastern part of the depression occupied by moraine-dammed Lake Low Boguty (water edge about 2387 m a.s.l.) in the proximal part of the debris cone of a short right tributary of the Boguty river. Repeated debris flows passed along this tributary valley during the Holocene [5]. Wash out and erosion of the eastern lake shore resulted in natural outcrops formation. One of these outcrops, about 70 cm high, was studied using different methods.

## 3. Methods

Pedostratigraphy and field morphology of peats and underlying lacustrine deposits were described in a shore-lake outcrop; pollen analysis and complex group analysis of biological composition were carried out for reconstructing paleoenvironmental conditions and sedimentation patterns.

Samples for pollen analysis were prepared applying the standard procedure [6]. Pollen and spores were identified using a microscope with 400x magnification. Microphotography of preparations was performed with a CarlZeiss Axioskop. Pollen percentages were calculated based on the tree and herbs pollen sum. Registering spores and other non-pollen palynomorphes during pollen analysis widen paleoecological abilities of the method. Thus, the registered stomata give an additional information on distribution of tree vegetation in the immediate vicinity of the section; finds of fungi (Glomus) are indirect indicators of enhanced drainage (bog – meadow transition).

Method of group analysis of biological composition was initially developed for lacustrine deposits (gyttja) [7] and later was supplemented and successfully approved for other water-related environments [8]. Water suspension of native sample is analyzed under microscope (X280-400); all encountered remnants of flora and fauna (fragments of vascular plants, cists and skeletons of algae, sponge spicules etc.) are identified up to the lowest possible taxonomic level.

The species composition of forest vegetation from charcoal fragments was determined by V.S. Myglan, Siberian Federal University, Krasnoyarsk, Russia.

Litho-stratigraphic and morphological pedogenetic descriptions were provided following FAO guidelines for soil description [9]. The presented profile within the studied peatbog was classified in terms of World Reference Base for Soil Resources [10].

Detailed geomorphological investigations and process analyses were based on interpretation of aerial photographs, Landsat-TM images, topographic maps (1:50000), and field investigations including mapping of landforms and deposits of different genesis.

The selected exposures were studied to examine the sediments and landforms associations. Suitable material from key locations was sampled for determining the radiocarbon chronology of the climate changes and landscape evolution.

The radiocarbon ages were measured at the Cenozoic Geochronology Center SB RAS, Novosibirsk, and at the Institute of Geography RAS, Moscow, applying the standard method [11]. The conventional radiocarbon ages were calibrated (2-sigma standard deviation) applying the CALIB Rev 7.0.4 program [12], with the IntCal13 calibration data set [13].

## 4. Results

4.1. *Profile horizontation and morphology*. Hummocky permafrosted boggy meadow (66 cm of peat layers) was described under mesic herbaceous sedge-grass vegetation (Poaceae sp., Carex sp., Gentiana tenella Rottb., Potentilla sericea, Dendranthema zawadskii (Herb) Tzvel, Paraquilegia microphylla (Royle) Drumm. Et Hutch). The terrain has naturally drained being recently cut by a new branch of a rivulet delta. Thus, the peat has run dried and started to degradate. The profile includes 7 horizons. The upper one is humified sodded peat. It is followed by five peaty horizons varying in color, texture and structure (or its absence), in degree of humification, in proportion of mineral admixture. Lacustrine silty loam underlies peat (figure 2).

Following horizons were described: **1)** Ha1 0-12 cm strongly decomposed peat (sapric material), with a considerable share of minerogenic components (sandy loam), moist, 7,5YR 2/2 brownish black, weak subangular blocky structure, root-interlaced sod; **2)** H2a 12-19cm an alternation of 1 cm interlayers: strongly decomposed (hemic) 7,5 YR 4/2 grayish brown and moderately decomposed (fibric), 7,5 YR 3/3 dark brown. Both contain little admixture of fine grained sand. Moist. **3)** H3a@ 19-30(34) cm moderate to strongly humified peat, cryoturbated, slightly moist to moist, 7,5YR 3/4 dark brown weak mostly subangular blocky structure; **4)** H4a,@ 30 (34)-47 cm very strongly humified, slightly moist, 7,5 YR 2/3 very dark brown. It is very loose and comparatively well structured; the structure is subangular-blocky; **5)** H5 47-55 cm weakly humified fibric peat, laminated, compressed, moist, 5YR4/4 dull reddish brown; **6)** H6@ 55-58(66) cm moderately to weakly humified peat, slightly moist, 7,5YR 2/3 very dark brown, loose, earthy; turbated: with small wedges; **7)** Lg 58(66) cm and deeper lacustrine silty loam, wet, 5Y 4/1 gray, 4/2 grayish olive, schlieren texture, moderately developed granular cryogenic structure, thixotropic. High ice content permafrost occurs below 71 cm.

**Figure 2.** Hummocky permafrosted boggy meadow near Lake Low Boguty. Photo (a) and studied section (b).

4.2. *Group analysis of biological composition.* The group analysis of biological composition conducted for 7 peat samples has allowed getting data on ecology of a local lacustrine-boggy system and to discriminate four evolutionary phases since the peatbog formation within the former lake bottom (figure 2). This was a result of cutting a moraine dam and dropping the level of Lake Low Boguty earlier 7600 BP.

**Phase 1:** The peaty material at the depth of 50 – 65 cm (**horizons 5, and 6**) was accumulated in a shallow lake within in-shore zone overgrown with water-bog plants and hypnum mosses. They gave refuge to a wide variety of diatoms (27-94% of the whole residues' sum) and yellow-green algae (*Chrysophytha* sp.) (1-1.5%). Attached forms prevail among the diatoms (about 93% of the diatoms' sum); *Fragilaria* (65%) and *Diatoma* (27%) are dominants. Representatives of benthic algae compose only 3%. A few fragments of sponges indicate a weak flowage in the lake in the relative time.

**Phase 2:** The peat at the depth of 36 – 47 cm (**horizon 4**) does not contain algae, water-bog plants appear in small amounts. This could be an evidence of an open water disappearance. Vascular plants' residues are represented mostly by roots of sedges. Dwarf birch (*Betula fruticosa* Pall) appears.

**Phase 3:** The peat at the depth 12-36 cm (**horizons 2, and 3**) was formed in conditions of better water supply. Water-bog plants got more abundant at 30-36 cm (**horizon 3**). Sums of diatoms and euglenales rise up to 8%. Residues of grasses (Poaceae) took place along with sedges (Cyperaceae).

**Phase 4:** The deposits in the upper 6 cm of the section (**upper part of the horizon 1**) were accumulated under sedge-grass-herbaceous boggy meadow. Diatoms have nearly disappeared. Half of those which still occur is typical for boggy environment (*Pinnularia*), others (*Hantzschia amphioxis*) can inhabit soils.

4.3. *Pollen analysis.* In total, 12 samples from silty loam (2) and peat (10) deposits were collected for the analysis of palynomorphs. Four pollen complexes were established (figure 3). Their boundaries were determined by changing of dominant taxa. The lower complex (1) characterizes landscapes and vegetation during lacustrine sedimentation, all other complexes (2-4) – evolution of regional and local vegetation during subaerial peat accumulation. The description below is given from bottom to top.

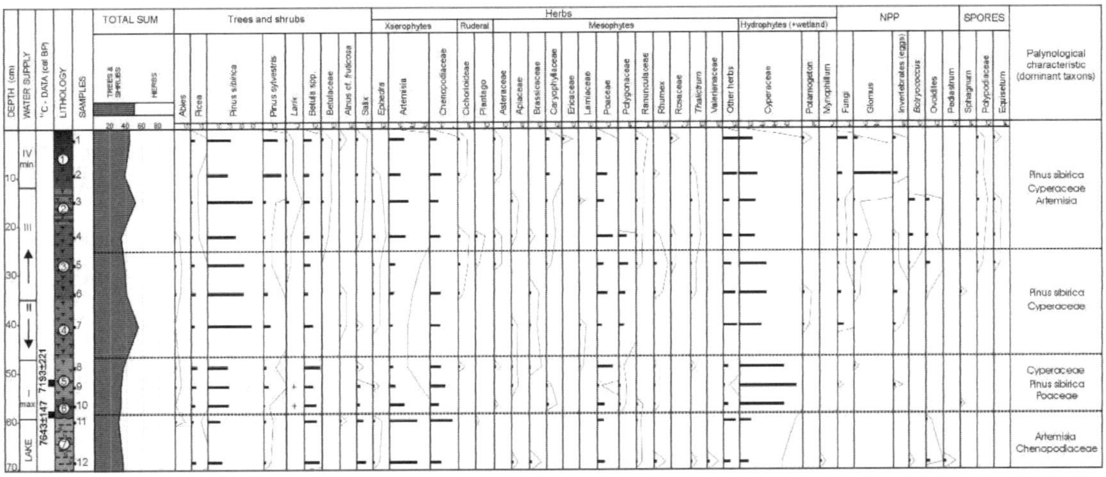


**Figure 3.** Pollen diagram,lithology and sample position in the studied section.

**Pollen complex 1:** olive-grey lacustrine clay with organic lenses at the depth 58 – 70 cm (samples 11, 12, **horizon 7**). The spectra of this zone are characterized by high contents of various herbs pollen (>60%). The dominance of *Artemisia* spp. and Chenopodiaceae is typical for pollen complex 1. Pollen percentage of trees and shrubs is about 30-35% and consist of *Betula* spp., *Pinus* spp., *Pinus sibirica* Du Tour, *Salix*, *Picea*. Percentage of microphytoplankton is low and presented by *Pediastrum*, *Spirogyra*, Zygnemataceae and *Botryococcus*.

**Pollen complex 2:** Peat of **horizons 5 and 6** (48 – 58 cm, samples 8 – 10). This complex is dominated by herbs pollen too. Among the herbaceous taxa the contribution of Cyperaceae increases. Trees and shrubs pollen are presented by *Pinus sibirica, Betula, Picea, Pinus sylvestris* L. The coniferous stomata were found in the samples 9 and 10 (figure 4).

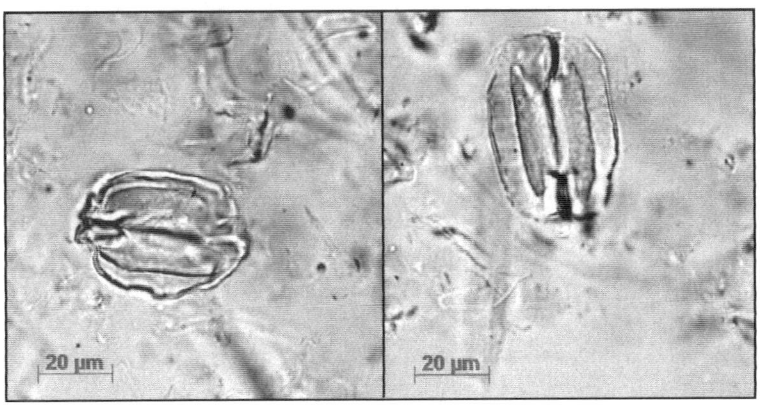


**Figure 4.** Coniferous stomata discovered in the samples 9 and 10.

**Pollen complex 3:** Peat of **horizons 3 and 4** (25 – 48 cm, samples 7 – 5). The complex is characterized by dominance of Cyperaceae and *Pinus sibirica* pollen. Non-pollen palynomorphs are also presented in spectra (fungal spores and hyphae, *Glomus* particulary, eggs of invertebrate). Pollen grains of *Larix* were determined in the sample 5, as well as few remains of algae – Spirogyra and Botryococcus.

**Pollen complex 4:** Peat, decomposed peat with minerogenic components, and humus of **horizons 1, 2, and 3** (0 – 25 cm (samples. 4 – 1)). The uppermost spectrum of the complex 4 is characterized by an approximately equal amount of trees and herbs pollen. Among the herbaceous taxa the contribution of *Artemisia* increases (up to 20%). *Larix* pollen is presented in all samples of this interval.

*4.4. Radiocarbon analysis*

Radiocarbon ages of two lower peat horizons at 57 – 67 and 50 – 57 cm were obtained – 7643±147 cal BP (IGAN 4823) and 7193±221 cal BP (IGAN 4825) respectively. Sixteen more radiocarbon ages of organic material from deposits of various geneses within the Boguty basin [5, 14] cover time interval 14000 – 300 cal BP and suggest some chronological bench marks of climatically driven landscape evolution in the region.

## 5. Reconstructions of the local lacustrine-boggy system transformation, climate changes and landscape evolution

Multidisciplinary investigations in the Boguty basin with the main focus on surroundings of Lake Low Boguty allowed reconstructing paleoenvironmental conditions, which were preceded the peatbog formation. The age of paleopeat with fragments of tree and bush vegetation discovered in the head of the tributary valley near Lake Low Boguty argues for complete degradation of Sartan glaciation or significant glaciers retreat (above 2500 m a.s.l., the altitude of paleopeat location) within the eastern periphery of the Chuya intermountain depression at about 14000 BP [14]. The fossil forest soils of about 11000 BP and charcoal (*Larix*) fragments (8500-7800 BP) indicate the prolonged period of developing forest vegetation in this part of the Boguty basin. Thus, it could be stated that the climate in the first half of the Holocene in the SE Altai was warmer and more humid in comparison with the modern one. Forest vegetation occupied contemporarily treeless areas, including slopes of the Boguty depression.

Accumulation of lacustrine loams at the bottom of the studied peatbog took place before 7600 BP. Freshwater algae *Pediastrum, Spirogyra* and *Botryococcus* settled in the lake at that time. The plant communities were dominated by *Artemisia* spp. and Chenopodiaceae. The lake was more extensive (up to 2400 m a.s.l.) and most likely it was surrounded by open landscapes, despite the presence of tree and shrubby vegetation (up to 30-35% in pollen spectra) - *Betula* spp., *Pinus* spp., and single *Pinus sibirica, Salix, Picea* (figure 3). It is interesting that *Larix* pollen was not found in the analyzed samples of lacustrine deposits, although fragments of larch charcoals with the age of 8500 – 7800 BP were collected from buried soils not more than 1 km upstream the tributary valley from the studied section. The absence of *Larix* pollen could be explained by both local distribution of larches or erosion (washing out) the upper part of the lacustrine deposits in the result of subsequent lowering of the lake level.

After the lake level lowering, which was associated with cutting the moraine dam, within the lake littoral, settled by higher water-bog plants and hypnum mosses with a diverse diatom flora, two lower peat horizons were developing about 7600 – 7200 BP. The presence of coniferous stomata in these horizons is an evidence of growing tree vegetation near the section at that time. The episode of significant lowering of the water edge, occurred later than 7200 BP, is recorded in all proxy archives – in sedimentary record (horizon 4), in biological composition of plant remnants (phase 2), and is also displayed in pollen spectra by appearance of non-pollen palynomorphs (fungal spores and hyphae of *Glomus*, eggs of invertebrate).

The subsequent improvement in the water supply of the peatbog (horizons 3, 2; phase 3) leads to increasing the number of water-bog plants, diatomaceous, euglenales and green algae. However, already at this time the vegetation in the lake surrounding reflects the enhanced aridity. The pollen of xerophytes, indicators of steppe landscapes, begins to predominate in the pollen spectra. Finally, the upper horizon of the section corresponds to the stage of stable peatbog draining, and its transformation into boggy meadow. Further radiocarbon dating of upper peat horizons will allow correlating periods of enhanced aridity, revealed from peat deposits of the Boguty basin, with the same periods recorded in pedosedimentary records of the SE Altai and SW Tuva [14, 15].

Generally, observed fluctuations of water supply in the lacustrine-boggy system within the Boguty basin may be associated with the stadial advances of mountain glaciers in the second half of the Holocene established for the SE Altai [16, 17]. Palynological data indicate that during all stages of peatbog developing water and fen plants (sedges, reeds, rdest etc.) were representatives of local vegetation around and within the lake. Among herbaceous vegetation, both xerophytes (*Artemisia, Ephedra*, Chenopodiaceae) and mesophytic plants were widely distributed. Pollen of trees and shrubs (mainly coniferous and to a lesser degree birch and shrub alder) is significantly represented in pollen spectra and may indicate spreading of forest vegetation within the study area. This thesis is also supported by presence of coniferous stomata in two lowest peat horizons, as well as heavy *Larix* pollen in three upper horizons. Perhaps the tree vegetation in the last 7000 – 8000 years, until its complete vanishing, was only locally distributed in the Boguty basin. Buried forest soils and charcoals of younger age were not discovered within the study area.

Presented multidisciplinary investigations of peats in the Boguty river basin are in good agreement with the results obtained by Blyakharchuk et al [18] in the course of palynological analysis of cores from Grusha and Akhol Lakes, located within neighboring areas of the SW Tuva at the foot of the opposite slope of the Chikhachev range.

**Acknowledgment**
The study was supported by state assignment project (no. 0330-2016-0015) and partly funded by Russian Foundation for Basic Research (project no. 16-05-01035). Pedostratigraphic studies and partly field researches were also supported by the Russian Science Foundation (project no. 14-27-00133).

**References**
[1] Devyatkin E V 1965 *Cenozoic deposits and neotectonics of Southeastern Altai* (Moscow: USSR Academy of Science) p 244
[2] Butvilovsky V V 1993 *Paleogeography of the Last Glaciation and the Holocene of Altai: a Catastrophic Events Model* (Tomsk: Tomsk University Press) p 253
[3] Agatova A R, Nepop R K, Slyusarenko I Y, 2017 *Archaeology, Ethnology & Anthropology of Eurasia* **45/1** 25–35
[4] Ganiushkin D, Chistyakov K and Kunaeva E 2015 *Environmental Earth Sciences* **74 (3)** 1883–904
[5] Nepop R K, Agatova A R 2017 *Advancing Culture of Living with Landslides* **5** 389–94
[6] Faegri K, Iversen J, Krzywinski K and Kaland P E 1989 *Textbook of pollen analysis* (Chichester-New York-Toronto-Sinapore: John Willey & sons) p 328
[7] Korde N V 1960 *Biostratigraphy and Typology of Russian Gyttja* (Moscow: USSR Academy of Science) p 220
[8] Uspenskaya O N 1986 *General patterns of formation and development of lakes. Methods of studying the history of lakes*, ed A F Treshnikov, V A Kvasov and V A Rumjancev (Leningrad: Nauka) pp 146–51
[9] Jahn R, Blume H P, Asio V B, Spaargaren O, Schad P 2006 *Guidelines for soil description* (FAO)
[10] IUSS Working Group WRB 2007 World Reference Base for Soil Resources 2006, first update 2007. *World Soil Resources Reports* **103**
[11] Arslanov A A 1987 *Radiocarbon: Geochemistry and Geochronology* (Leningrad: Leningrad State University Press) p 300
[12] Stuiver M and Reimer P J 1993 1993 *Radiocarbon* **35 (1)** 215–230
[13] Reimer P J, Baillie M G, Bard E, Bayliss A, Beck J W, Blackwell P G, Ramsey C B, Buck C E, Burr G S, Edwards R L and Friedrich M 2009 *Radiocarbon* **51 (4)** 1111–50
[14] Agatova A R, Nepop R K, Bronnikova M A, Slyusarenko I Yu and Orlova L A 2016

*Archaeological and Anthropological Sciences* **8** 419–40
[15] Bronnikova M A, Konoplianikova Yu V, Agatova A R, Zazovskaya E P, Lebedeva M P, Turova I V, Nepop R K, Shorkunov I G, Cherkinsky A E 2017 *Eurasian Soil Science* **50 (2)** 142–57
[16] Agatova A R, Nazarov A N, Nepop R K and Rodnight H 2012 *Quaternary Science Reviews* **43** 74–93
[17] Agatova A R, Nepop R K, Slyusarenko I Yu, Myglan V S, Nazarov A N and Barinov V V 2014 *Quaternary International* **324** 6–19
[18] Blyakharchuk T, Wright H, Borodavko P, Van der Knaap W O and Ammann B 2007 *Palaeogeography, Palaeoclimatology, Palaeoecology* **245** 518–534

# Trees as methane sources: A case study of West Siberian South taiga

**A I Churkina[1,2], S Yu Mochenov[1], S F Sabrekov[2,3], M V Glagolev*[1,2,3,4], D V Il'yasov[2,4], I E Terentieva and S S Maksyutov[5]**

[1] Faculty of Soil Science, Moscow State University, 1 Leninskie gory, Moscow 119991, Russia
[2] Water Problems Institute of RAS, 3 Gubkina, Moscow 119991, Russia
[3] UNESCO Department "Environmental Dynamics and Global Climate Changes", Yugra State University, Khanty-Mansiysk, 628012, Russia
[4] Institute of Forest Science Russian Academy of Sciences, Moscow Region, Russia
[5] Center for Global Environmental Research, National Institute for Environmental Studies, Nishi Odori Street, Tsukuba 305-8506, Japan

*Email: m_glagolev@mail.ru

**Abstract.** Within this study, we were measuring methane emission from the tree trunks, leaves and branches in the seasonally flooded forest and in the forested bogs (pine-shrub-sphagnum ecosystems or "ryams") in south taiga zone of Western Siberia. Our results suggest that the tree trunks may act as a methane conductor from the soil to the atmosphere bypassing the methanotrophically active zones of soil. The tree methane flux depends on a trunk diameter and an ecosystem type. The average methane emission from tree trunks was 0.0061±0.0003 mg $CH_4 \cdot m^{-2} \cdot h^{-1}$ per unit of ground area. The methane emission from branches and leaves was zero.

## 1. Introduction

High-quality quantitative estimates of the $CH_4$ budget are crucial for predicting climate changes, managing Earth's carbon reservoirs, and understanding atmospheric chemistry [1]. The growth of the methane concentration reached up to 1% per year in the XX century. Currently several methane sources are supposed to contribute to the emission and the wetlands are the largest natural methane source [2].

However, there are some differences in $CH_4$ emission estimates by bottom-up models using direct flux measurements and by top-down models using the inverse modeling of tall tower and satellite data [3]. It is suggested that we may still be missing some unknown methane source.

The majority of chamber flux measurements were made from soil and excluded trees [4, 5, 6, 7]. At the same time the role of plant-mediated methane transport from soil to the atmosphere has been well known for a long time [8, 9, 10, 11]. Now we have a number of evidences that methane may be emitted out of and through trees [12, 13, 14, 15]. In this paper, we studied trees in seasonally flooded forests potentially significant methane sources in Western Siberia.

The tasks of this study were to make field measurements of methane emitted out of and through tree trunks and leaves and to study the environmental controls of the methane fluxes from trees (tree

IOP Conf. Series: Earth and Environmental Science **138** (2018) 012002 doi:10.1088/1755-1315/138/1/012002

diameter, chamber installation height, temperature, daytime) in various *Betula pendula* and *Pinus sylvestris* ecosystems.

## 2. Materials and methods

*2.1 Research sites*

The $CH_4$ flux measurements from the trees and soil were carried out in Bakcharsky region of Tomsk oblast in the south taiga zone of Western Siberia during July and August 2016-2017 (figure 1). The measurements were conducted at three sites in seasonally flooded forest and two sites in forested bogs (pine-shrub-sphagnum ecosystems or "ryams"). The site at seasonally flooded forest (Wet Forest_2) was chosen due to it's closeness to the Bakcharskoe bog. The site at seasonally flooded forest within the drained bog (site Wet Forest_3) was chosen to compare the methane emission from trees at sites with different distance from the bog and different WTL (Water table level). The site at seasonally flooded forest within the floodplain of Bakchar river tributary (site Wet Forest_Polynyanka) was chosen to compare the tree methane emissi different relief. The measurements at ryams (sites Low ryam_Bakcharskoe bog_1, 2) were made to understand their contribution in methane emission in this ecosystem and to compare with the tree methane emission in seasonally flooded forests.

**Figure 1.** Test sites.

The detailed site description is given in the table 1. The test sites are the same sites which were mentioned in the article Mochenov et al. published in this Proceedings Book.

**Table 1.** Description of test sites.

| Point | Coordinates | | Characteristics | | |
|---|---|---|---|---|---|
| | Latitude | Longitude | pH | EC | The most abundant woody species [a)] |
| | | Wet Forest_Polynanka | | | |
| fp_2 | 56.96189 | 82.51528 | 6.89-7.280 | 680-901 | B.p., C.v. |
| fp_3 | 56.96181 | 82.51528 | 7.01-7.16 | 523-623 | B.p., C.v. |
| fp_6 | 56.96128 | 82.51533 | 6.69-7.31 | 508-809 | B.p., C.v. |
| fp_7 | 56.96131 | 82.51536 | 7.00-7.62 | 508-809 | B.p., C.v. |
| fp_8 | 56.96125 | 82.51542 | 6.91-7.20 | 855-880 | B.p., F.r., A.p. |
| fp_9 | 56.96128 | 82.51544 | n.d.[b)] | n.d.[b)] | B.p., A.p., C.a. |
| | | Wet Forest_2 | | | |
| Tr. PWF_2 | 56.83128 | 82.85150 | n.d.[b)] | n.d.[b)] | B.p. |
| | | Wet Forest_3 | | | |
| WF_BD_01 | 56.87308 | 82.86350 | 5.03-5.17 | 44-78 | P.s., B.p. |
| WF_BD_1 | 56.87317 | 82.86386 | 5.31-6.26 | 101-390 | P.s. |
| WF_BD_5 | 56.87314 | 82.86381 | 4.74-5.50 | 64-131 | P.s. |
| WF_BD_7 | 56.87319 | 82.86367 | 4.27-4.66 | 130-143 | P.s. |
| WF_BD_8 | 56.87322 | 82.86358 | 4.29-5.48 | 86-157 | P.s. |
| | | Low ryam_Bakcharskoe bog_1 | | | |
| Ryam_1 | 56.85467 | 82.85156 | 3.99-4.13 | 62-64 | P.s. |
| | | Low ryam_Bakcharskoe bog_2 | | | |
| R_1 | 56.87217 | 82.84978 | 3.10-4.22 | 76-101 | P.s. |
| R_2 | 56.87206 | 82.84981 | 3.83-3.83 | 106-106 | P.s. |
| R_3 | 56.87206 | 82.84983 | 3.80-3.94 | 186-205 | P.s. |

[a] C.c. - *Chamaedaphne calyculata*, An.p. - *Andromeda polifolia*, S.s. - *Sphagnum* sp., B.p. - *Betula pendula*, C.v. - *Carex vesicaria*, F.r. - *Festuca rubra*, A.p. - *Aegopodium podagraria*, C.a. - *Calamagrostis arundinacea*, P.s. - *Pinus sylvestris*
[b] WTL was deeper than 60 cm.

*2.2 Methods*

The methane flux measurements from tree trunks were made in accordance with a chamber method modification by Siegenthaler et al. [16]. In our case, the chamber height was 2.5 cm, its volume was one liter, and the chamber walls were joined together using the universal glue "Moment" (Henkel, Germany). The gas sample volume was 5-6 ml. The exposure time varied from 8 minutes to 2 hours. The median error of individual measurements from the tree trunk was 0.008 $mgCH_4 \cdot m^{-2} \cdot h^{-1}$.

To measure methane emitted out of leaves and branches, the ends of branches were placed into 1-5 liter plastic bottles sealed by k-flex stopper in place of a contact. To take 5 ml samples through the k-flex stopper, we used a silicone tube with inner diameter 2 mm and 12 ml syringes. Three samples were taken out when the 5-liter bottle was used and one sample in a case of the 1-liter bottle. The methane flux for the 5-liter bottle measurements were calculated by linear regression with the weights and by non-linear (exponential) regression for emission and consumption, respectively, using chamber headspace concentrations versus measuring time [17]. The methane flux for the 1-liter bottle measurements was calculated by averaging two samples from the same branch of a tree.

The methane and carbon dioxide flux measurements from the soils were made by classical chamber method (as was described in [18]) in parallel with the methane flux measurements from trees. The chambers were installed close to the measured trees.

The concentrations were analyzed no later than 18 hours after gas sampling (syringes with stoppers were stored in NaCl concentrated solution before the analysis). Gas concentrations were measured using a modified gas chromatograph KhPM-4 ("Chromatograph", Russia) with a flame ionization

IOP Conf. Series: Earth and Environmental Science **138** (2018) 012002 doi:10.1088/1755-1315/138/1/012002

detector from a chromatograph LHM-80 ("Chromatograph", Russia) with 1 m stainless steel column (2.5 mm o.d.) filled with Sovpol (80-100 mesh) at 35°C with hydrogen as a carrier gas (flow rate 5 ml/min) and the loop volume of 0.5 ml. For chromatograph calibration, we used methane and air standard mix with $CH_4$ 1.99, 5.08 and 9.84 ppm concentrations of ±0.01 ppm accuracy (National Institute for Environmental Studies, Japan). The volume of injected sample was 2-3 ml. Every gas sample from a syringe was analyzed in 3-4 replicates; the average concentration was used for the flux calculation.

The methane flux from trees was calculated using the difference between the atmosphere gas concentration before the measurements and the final $CH_4$ concentration in the chamber headspace. The fluxes in the table 2 were calculated by averaging from 2 to 3 replications of individual flux measurements from the same place of a tree.

To analyze the data, we used Wilcoxon rank sum test in MATLAB R2009a, regression analysis and the zero comparison test using t-criteria [19]. The statistical significance P-value was 0.05.

## 3. Results

The methane fluxes measured from leaves and branches were too small (comparing to the measurement error) if they actually emit the methane (table 2). The zero comparison test showed that leaves and branches methane fluxes were indistinguishable from zero with 0.05 significance level. The methane fluxes from tree trunks were higher, and the zero comparison test indicated that fluxes could be distinguished from zero with 0.05 significance level. The results of the individual methane flux measurements from tree trunks are given in the table 3.

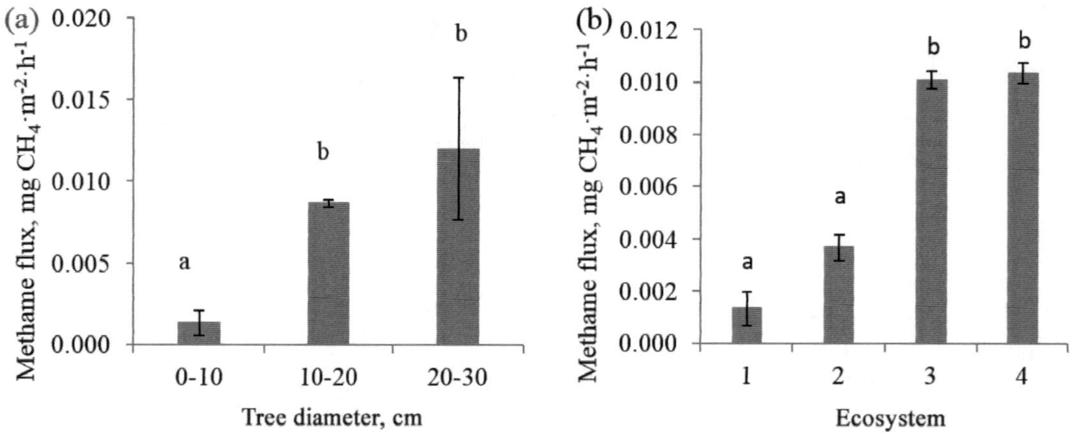


| n | 11 | 13 | 11 |
|---|---|---|---|

| n | 11 | 4 | 12 | 8 |
|---|---|---|---|---|

**Figure 2.** Methane flux medians from tree trunks: (a) dependence on the tree diameter, (b) dependence on the ecosystem type: «1» – Low ryam_Bakcharskoe bog_1, 2; «2» – Wet Forest_2; «3» – Wet Forest_3; «4» – Wet Forest_Polynyanka. The medians that differed by the Wilcoxon rank sum test was marked with letters «a» and «b». The specified error is a weighted median deviation. The «n» is a number of measured fluxes.

IOP Conf. Series: Earth and Environmental Science **138** (2018) 012002 doi:10.1088/1755-1315/138/1/012002

**Table 2.** Leaves and branches $CH_4$ fluxes, mg $CH_4 \cdot m^{-2} \cdot h^{-1}$.

| Point | Date | Time | Exposure time | Flux from leaves[a] | Error of flux from leaves[a] | Flux from branches | Error of flux from branches[b] | Total flux [b] | Error of total flux | Error code[d] | Air temperature, °C | Species[e] | Chamber volume, l |
|---|---|---|---|---|---|---|---|---|---|---|---|---|---|
| Low ryam_Bakcharskoe bog_1 | | | | | | | | | | | | | |
| Ryam_1 | 17.8 | 18:08 | 0:09:42 | -0.027 | 0.03 | -0.217 | 0.26 | -0.396 | 0.50 | 1 | 23.5 | *P.syl.* | 1 |
| Ryam_1 | 21.8 | 16:25 | 0:28:26 | 0.000 | 0.004 | -0.001 | 0.075 | 0.000 | 0.004 | 95 | 17.7 | *P.syl.* | 5 |
| Wet Forest_2 | | | | | | | | | | | | | |
| Tr. PWF_2 | 21.8 | 13:30 | 0:32:18 | -0.002 | 0.005 | -0.040 | 0.077 | -0.002 | 0.005 | 1 | 17.7 | *P.sib.* | 1 |
| Tr. PWF_2 | 21.8 | 13:41 | 0:56:53 | 0.001 | 0.007 | 0.013 | 0.075 | 0.001 | 0.007 | 6 | 17.6 | *P.o.* | 5 |
| Tr. PWF_2 | 21.8 | 13:49 | 0:55:20 | 0.004 | 0.008 | 0.081 | 0.132 | 0.004 | 0.007 | 4 | 17.5 | *B.p.* | 5 |
| Wet Forest_3 | | | | | | | | | | | | | |
| WF_BD_8 | 19.8 | 15:07 | 2:21:47 | 0.001 | 0.005 | 0.024 | 0.072 | 0.001 | 0.004 | 6 | 20.0 | *P.syl.* | 5 |
| WF_BD_1 | 19.8 | 14:32 | 0:43:37 | 0.001 | 0.001 | 0.019 | 0.012 | 0.001 | 0.001 | 1 | 20.0 | *P.syl.* | 1 |
| WF_BD_1 | 19.8 | 15:25 | 2:05:58 | 0.003 | 0.017 | 0.028 | 0.204 | 0.003 | 0.016 | 6 | 20.0 | *B.p.* | 5 |
| WF_BD_8 | 19.8 | 12:54 | 1:19:00 | 0.011 | 0.035 | 0.025 | 0.083 | 0.008 | 0.024 | 6 | 19.1 | *P.sib.* | 5 |
| WF_BD_8 | 19.8 | 13:39 | 1:04:23 | 0.005 | 0.019 | 0.080 | 0.271 | 0.005 | 0.017 | 6 | 19.6 | *P.syl.* | 5 |
| WF_BD_8 | 19.8 | 14:58 | 2:29:27 | -0.001 | 0.051 | -0.001 | 0.123 | 0.000 | 0.036 | 95 | 20.0 | *P.sib.* | 5 |
| WF_BD_1 | 19.8 | 13:23 | 0:54:38 | 0.003 | 0.017 | 0.024 | 0.201 | 0.001 | 0.016 | 6 | 18.9 | *B.p.* | 5 |
| WF_BD_1 | 19.8 | 17:54 | 0:21:57 | 0.013 | 0.028 | 0.185 | 0.387 | 0.012 | 0.027 | 6 | 17.1 | *B.p.* | 5 |
| WF_BD_1 | 19.8 | 18:48 | 0:10:51 | -0.003 | 0.031 | -0.006 | 0.065 | -0.002 | 0.022 | 1 | 16.9 | *P.sib.* | 1 |
| WF_BD_8 | 19.8 | 19:31 | 0:19:46 | 0.024 | 0.091 | 0.059 | 0.221 | 0.017 | 0.064 | 6 | 15.8 | *P.sib.* | 5 |
| WF_BD_8 | 19.8 | 19:14 | 0:22:27 | -0.008 | 0.200 | -0.121 | 2.987 | -0.008 | 0.187 | 95 | 16.5 | *P.syl.* | 5 |

[a] if assumed that all methane is produced by leaves;
[b] if assumed that all methane is produced by branches;
[c] if assumed that all methane is produced by both leaves and branches;
[d] error code: 1 – standard deviation, 4 - standard error, 6 - error derived from solving linear programming problem, 95 - confidence interval is 95%;
[e] - *P.syl.* - *Pinus sylvestris*, *P.sib.* - *Pinus sibirica*, *P.o.*- *Picea obovata*, *B.p.* - *Betula pendula*.

**Table 3.** Tree $CH_4$ fluxes, mg $CH_4 \cdot m^{-2} \cdot h^{-1}$.

| Point | Date | Day time | Exposure time | $CH_4$ flux | Error[a] | Chamber installation height, cm | Tree diameter, cm | Air temperature, °C | Species[b] |
|---|---|---|---|---|---|---|---|---|---|
| Low ryam_Bakcharskoe bog_1, 2 | | | | | | | | | |
| Ryam_1 | 10.8 | 13:17 | 1:01:40 | 0.000 | 0.003 | 0 | 25 | 20.1 | *P.s.* |
| Ryam_1 | 10.8 | 14:11 | 1:04:06 | 0.001 | 0.004 | 50 | 23 | 20.0 | *P.s.* |
| Ryam_1 | 10.8 | 13:10 | 1:03:30 | 0.052 | 0.068 | 150 | 20 | 20.4 | *P.s.* |
| Ryam_1 | 21.8 | 16:09 | 0:41:43 | 0.009 | n.d.[b] | 250 | 5 | 18.0 | *P.s.* |
| Ryam_1 | 21.8 | 16:06 | 0:43:43 | 0.004 | n.d.[b] | 50 | 6 | 18.0 | *P.s.* |
| R_1 | 13.8 | 17:11 | 0:22:34 | -0.001 | 0.014 | 150 | 5 | 16.9 | *P.s.* |
| R_2 | 13.8 | 17:00 | 0:14:13 | 0.001 | 0.025 | 50 | 5 | 19.7 | *P.s.* |
| R_3 | 13.8 | 17:02 | 0:18:21 | 0.007 | 0.012 | 50 | 9 | 18.3 | *P.s.* |
| R_1 | 13.8 | 18:25 | 0:07:42 | -0.005 | 0.008 | 150 | 5 | 16.3 | *P.s.* |
| R_3 | 13.8 | 16:32 | 0:15:28 | -0.019 | 0.070 | 0 | 9 | 19.0 | *P.s.* |
| R_3 | 13.8 | 16:09 | 0:19:05 | -0.019 | 0.023 | 200 | 9 | 18.1 | *P.s.* |
| R_2 | 13.8 | 15:50 | 0:18:46 | 0.009 | 0.020 | 200 | 4 | 18.9 | *P.s.* |
| R_2 | 13.8 | 14:47 | 0:37:25 | 0.005 | 0.004 | 0 | 5 | 18.6 | *P.s.* |
| Wet Forest_2 | | | | | | | | | |
| Tr. PWF_2 | 21.8 | 13:25 | 1:05:40 | 0.004 | 0.002 | 150 | 14 | 18.3 | *P.s.* |
| Tr. PWF_2 | 21.8 | 13:45 | 0:56:23 | 0.005 | 0.001 | 150 | 14 | 17.9 | *P.s.* |
| Tr. PWF_2 | 21.8 | 13:51 | 0:55:09 | 0.000 | 0.006 | 150 | 5 | 17.9 | *B.p.* |
| Tr. PWF_2 | 21.8 | 13:59 | 0:53:42 | 0.003 | 0.005 | 0 | 6 | 17.8 | *B.p.* |
| Wet Forest_3 | | | | | | | | | |
| WF_BD_5 | 4.8 | 14:55 | 0:15:39 | 0.219 | n.d.[b] | n.d.[b] | n.d.[b] | 28.3 | *P.s.* |
| WF_BD_01 | 4.8 | 14:54 | 0:26:34 | 0.023 | n.d.[b] | n.d.[b] | n.d.[b] | 28.5 | *B.p.* |
| WF_BD_7 | 6.8 | 13:44 | 0:31:10 | 0.012 | 0.019 | 0 | 20 | 27.9 | *P.s.* |
| WF_BD_7 | 6.8 | 14:39 | 0:21:31 | 0.012 | 0.002 | 50 | 20 | 29.0 | *P.s.* |
| WF_BD_7 | 6.8 | 15:18 | 0:33:30 | -0.001 | 0.004 | 50 | 15 | 28.2 | *B.p.* |
| WF_BD_7 | 6.8 | 15:56 | 0:22:33 | 0.043 | 0.078 | 150 | 18 | 28.7 | *P.s.* |
| WF_BD_7 | 19.8 | 17:56 | 1:02:13 | 0.010 | 0.003 | 200 | 14 | 17.7 | *P.s.* |
| WF_BD_7 | 19.8 | 16:19 | 0:21:36 | 0.018 | 0.007 | 200 | 14 | 20.1 | *P.s.* |
| WF_BD_7 | 19.8 | 18:09 | 1:04:52 | 0.001 | 0.002 | 200 | 14 | 17.2 | *P.s.* |
| WF_BD_7 | 19.8 | 14:13 | 0:29:46 | 0.071 | 0.040 | 0 | 19 | 19.9 | *P.s.* |
| WF_BD_7 | 19.8 | 17:50 | 1:01:43 | 0.003 | 0.004 | 200 | 15 | 17.9 | *P.s.* |
| WF_BD_7 | 19.8 | 14:24 | 0:23:36 | 0.009 | 0.005 | 50 | 16 | 19.9 | *P.s.* |
| WF_BD_7 | 19.8 | 16:10 | 0:22:53 | 0.010 | 0.004 | 200 | 15 | 20.4 | *P.s.* |
| WF_BD_7 | 19.8 | 17:06 | 2:03:01 | 0.003 | 0.005 | 200 | 3.5 | 18.2 | *B.p.* |
| Wet Forest_Polynyanka | | | | | | | | | |
| fp_3 | 1.8 | 16:42 | 0:21:11 | 0.688 | n.d.[b] | n.d.[b] | n.d.[b] | 28.7 | *B.p.* |
| fp_3 | 1.8 | 15:37 | 0:59:14 | 0.132 | n.d.[b] | n.d.[b] | n.d.[b] | 29.7 | *B.p.* |
| fp_2 | 1.8 | 15:58 | 2:31:36 | 0.580 | n.d.[b] | n.d.[b] | n.d.[b] | 27.7 | *B.p.* |
| fp_2 | 1.8 | 16:34 | 2:08:23 | 0.109 | n.d.[b] | n.d.[b] | n.d.[b] | 27.7 | *B.p.* |
| fp_6 | 9.8 | 15:18 | 0:20:03 | 0.008 | 0.009 | 200 | 28 | 17.7 | *B.p.* |
| fp_7 | 9.8 | 15:39 | 0:20:30 | 0.037 | 0.010 | 200 | 20 | 17.3 | *B.p.* |
| fp_7 | 9.8 | 17:34 | 0:20:00 | 0.141 | 0.068 | 100 | 20 | 14.6 | *B.p.* |
| fp_6 | 9.8 | 17:36 | 0:20:00 | 0.009 | 0.041 | 100 | 28 | 14.6 | *B.p.* |
| fp_9 | 9.8 | 15:21 | 0:20:57 | 0.003 | 0.025 | 200 | 20 | 17.6 | *B.p.* |
| fp_9 | 9.8 | 17:39 | 0:20:00 | 0.033 | 0.001 | 100 | 20 | 15.1 | *B.p.* |
| fp_8 | 9.8 | 15:24 | 0:20:00 | 0.003 | 0.033 | 200 | 10 | 17.6 | *B.p.* |
| fp_8 | 9.8 | 17:42 | 0:19:55 | 0.011 | 0.020 | 100 | 10 | 14.6 | *B.p.* |

[a] - error is a standard deviation from 2-3 measured flux values at the same point of a tree
[b] - *P.s.* - *Pinus sylvestris*, *B.p.* - *Betula pendula*
[c] - data from single test preliminary $CH_4$ flux measurements.

The linear regression analysis revealed that methane fluxes from tree trunks depended on a tree diameter (F=0.040, $R^2$=0.121; figure 2.a). The influence of chamber installation height, temperature, daytime and the exposure time on methane flux was not found though more measurements are needed for reliability.

The Wilcoxon rank sum test revealed that methane fluxes from *Betula pendula* and *Pinus sylvestris* trunks differed insignificantly (P=0.66) though more measurements are needed as well. The methane fluxes from tree trunks at ryams and seasonally flooded forest located nearby the Bakcharskoe bog also differed insignificantly (P=0.66). Insignificant variations in methane fluxes (P=0.73) were also observed among tree trunks of seasonally flooded forests within the drained bog and within the floodplain of the Bakchar river tributary (figure 2.b).

It was found that methane fluxes from tree trunks tended to be higher with the decrease of methane fluxes from soils (figure 3.a) and the increase of carbon dioxide fluxes (figure 3.b). However, detailed study is needed to confirm this dependence.

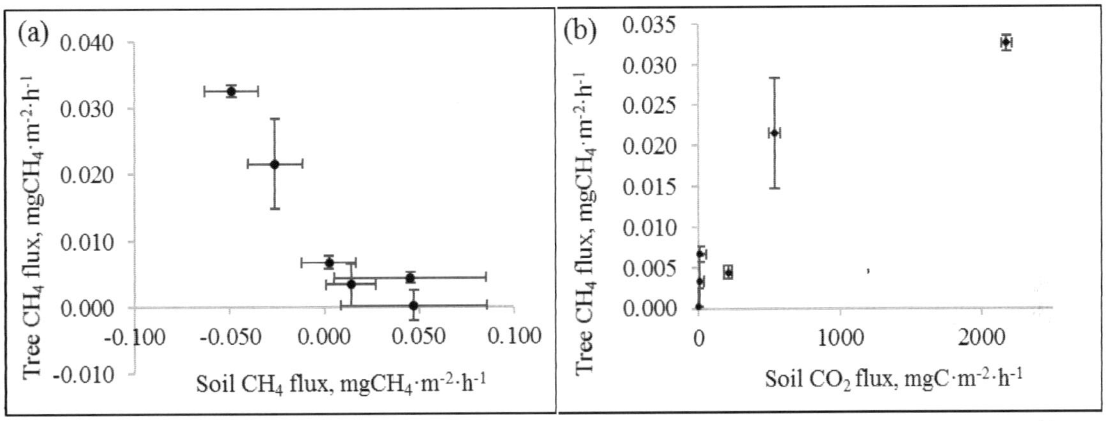


**Figure 3**. The relationship of $CH_4$ fluxes from trunks with $CH_4$ (a) and $CO_2$ (b) fluxes from soils.

## 4. Discussion

The increase of methane concentration in soils was found at 30 cm depth as shown in the article Mochenov et al. published in this Proceedings Book. As the tree roots consume water from deeper layers, they uptake water with high dissolved methane concentration. Peat soils absorb water more intensively, so plants get less water than in mineral soils. Therefore, the transpiration will be lower in peat soils leading to lower methane release through trees. This may explain methane flux differences at the sites (figure 2.b). The ryam and seasonally flooded forest near The Bakcharskoe bog is characterized by peat soils resulting in lower methane fluxes from trees at these sites. Soils in seasonally flooded forest at the drained bog are more mineral, while the floodplain soils contain no peat. Methane fluxes from trees at these sites are considerably higher in comparison to sites with peat soils.

The high methane concentration may also explain the more intensive methane fluxes from trees with a large trunk diameter as those trees have deeper and more extended roots, so they could uptake greater amount of water with dissolved methane. Pangala et al. [15] reported that methane fluxes decrease with the tree diameter raise. However, they measured methane fluxes from trees with a larger trunk diameter (20-130 cm) and the highest methane fluxes were produced by the trees with 20-50 cm diameter. Our data showed that the maximal fluxes were released by the trees with a 10-30 cm trunk diameter. More measurements are needed from larger trees, but they were rarely found in the studied ecosystems. The trees with a larger diameter may have different factors effecting $CH_4$ flux.

When the methane emits through the trees, the methane flux from soils may decrease due to one more methane release way diminishing the methane amount in the source. At the same time, the increase of $CO_2$ fluxes along with the methane flux increase from tree may be caused by microbiological activity.

The data from our study are mostly in a good correspondence with the data from the other studies (table 4). The highest $CH_4$ flux from trees was found at the tropical forested peatland [15] and the lowest $CH_4$ flux was found at the boreal coniferous forest [20]. The latitude effect on methane fluxes from trees might be caused by the plant transpiration increase with the temperature rise resulting in a greater methane release through the trees.

**Table 4.** Comparison of methane fluxes from trees by other studies.

| N | Tree $CH_4$ emission, mg $CH_4 \cdot m^{-2} \cdot h^{-1}$ | | Tree $CH_4$ emission, $mg \cdot m^{-2} \cdot h^{-1}$ (per unit of ground area) | Source |
|---|---|---|---|---|
| | min | max | | |
| 1 | 0.097 | 0.176 | 0.0027 | [13] |
| 2 | 0.0170±0.0014 | 0.185±0.007 | 0.1188±0.0142 | [15] |
| 4 | 0.0041±0.0008 | 0.101±0.022 | n.d. | [14] |
| 5 | 0.0000 | 0.0001 | 0.0001 | [20] |
| 6 | 0.0031 | 0.3600 | n.d. | [21] |
| 7 | 0.0833 | 0.8583 | n.d. | [22] |
| 8 | 0.0013±0.0018 | 0.0323±0.0009 | 0.0061±0.0003 | This study |

## 5. Conclusion

We may conclude that the tree trunks in periodically flooded forests may act as a methane conductor from the soil to the atmosphere bypassing the methanotrophically active zones of soil. The methane fluxes from lives and branches are too small and indistinguishable from zero while the methane fluxes from tree trunks are higher, distinguish from zero and depend on tree diameter and ecosystem type. There was found no dependence of methane fluxes from tree trunks and chamber installation height, temperature, daytime and tree species. The methane fluxes from trees varied from 0.0013±0.0018 to 0.0323±0.0009. The methane flux from trees per a unit of ground area was 0.0061±0.0003.

The positive correlation of methane fluxes from trees and carbon dioxide flux from soil were found while the correlation between methane fluxes from trees and methane flux from soil was negative.

## Acknowledgements

The study was supported by ArCS - Arctic Challenge for Sustainability Project. The authors express sincere gratitude to M.V. Chistotin (All-Russian Scientific and Research Institute of Agrochemistry named by D.N. Pryanishnikov) for providing the analiser DX6100-01, the specialists from V.E. Zuev Institute of Atmospheric Optics D.K. Davydov, A.V. Fofonov and M.Yu. Arshinov, who calibrated the instrument, I.S. Churkin for assistance with translation, A.A. Lebedev, who created the software, which allowed us to make the calculations of the average temperature and the weighted average much faster as well as M.V. Lukin and G.V. Uvarov for participating in the field measurements.

## References

[1] Mikaloff Fletcher S E, Tans P P, Bruhwiler L, Miller J B and Heimann M 2004 $CH_4$ sources estimated from atmospheric observations of $CH_4$ and its $^{13}C/^{12}C$ isotopic ratios: 1. Inverse modeling of source processes *Global Biogeochem. Cycles* **18** GB4004

[2] Andronova N G and Karol I L 1993 The contribution of USSR sources to global methane emission *Chemosphere* **26** 111–26

[3] Nisbet E G, Dlugokencky E J and Bousquet P 2014 Methane on the rise-again *Science* **343** 493–95

[4] Bubier J L, Moore T R and Juggins S 1995 Predicting Methane Emission from Bryophyte Distribution in Northern Canadian Peatlands *Ecology* **76** 677–93

[5] Dalva M, Moore T R, Arp P and Clair T A 2001 Methane and soil and plant community respiration from wetlands, Kejimkujik National Park, Nova Scotia: Measurements, predictions, and climatic change *J. Geophys. Res.* **106** 2955–62

[6] Glagolev M V, Sabrekov A F, Kleptsova I E, Filippov I V, Lapshina E D, Machida T and Maksyutov Sh Sh 2012 Methane Emission from Bogs in the Subtaiga of Western Siberia:

The Development of Standard Model *Eurasian Soil Sci.* **45** 947–57
[7] Sabrekov A F, Glagolev M V, Kleptsova I E, Machida T and Maksyutov S S 2013 Methane Emission from Mires of the West Siberian Taiga *Eurasian Soil Sci.* **46** 1182–93
[8] Dacey J W H and Klug M J 1979 Methane Efflux from Lake Sediments Through Water Lilies *Science* **203** 1253–55
[9] Sebacher D I, Harriss R C and Bartlett K B 1985 Methane Emissions to the Atmosphere Through Aquatic Plants *J. Environ. Qual.* **14** 40–46
[10] Whiting G J and Chanton J P 1992 Plant-dependent $CH_4$ emission in a subarctic Canadian fen *Global Biogeochem. Cycles* **6** 225–31
[11] Dorodnikov M, Knorr K.-H, Kuzyakov Y and Wilmking M 2011 Plant-mediated $CH_4$ transport and contribution of photosynthates to methanogenesis at a boreal mire: a $^{14}C$ pulse-labeling study *Biogeosciences* **8** 2365–75
[12] Zeikus J G and Ward J C 1974 Methane formation in living trees: a microbial origin *Science* **184** 1181-83
[13] Terazawa K, Ishizuka S, Sakata T, Yamada K and Takahashi M 2007 Methane emissions from stems of Fraxinus mandshurica var. japonica trees in a floodplain forest *Soil Biol. Biochem.* **39** 2689–92
[14] Gauci V, Gowing D J, Hornibrook E R, Davis J M and Dise N B 2010 Woody stem methane emission in mature wetland alder trees *Atmos. Environ.* **44** 2157–60
[15] Pangala S R, Moore S, Hornibrook E R and Gauci V 2012 Trees are major conduits for methane egress from tropical forested wetlands *New Phytol.* **197** 524–31
[16] Siegenthaler A, Welch B, Pangala S R, Peacock M and Gauci V 2016 Semi-rigid chambers for methane gas flux measurements on tree stems *Biogeosciences* **13** 1197–207
[17] Sabrekov A F, Glagolev M V, Alekseychik P K, Smolentsev B A, Terentieva I E, Krivenok L A and Maksyutov S S 2016 A process-based model of methane consumption by upland soils *Environ. Res. Lett.* **11** 075001
[18] Sabrekov A F, Glagolev M V, Fastovets I A, Smolentsev B A, Il'yasov D V, Maksyutov Sh Sh 2015 Relationship of Methane Consumption with the Respiration of Soil and Grass–Moss Layers in Forest Ecosystems of the Southern Taiga in Western Siberia *Eurasian Soil Sci.* **48** 841–51
[19] Tennant-Smith J 1985 *BASIC statistics* (London: Butterworths & Co)
[20] Machacova K, Bäck J, Vanhatalo A, Halmeenmäki E, Kolari P, Mammarella I, Pumpanen J, Acosta M, Urban O & Pihlatie M 2016 *Pinus sylvestris* as a missing source of nitrous oxide and methane in boreal forest *Sci. Rep.* **6** 23410
[21] Kutschera E, Khalil A, Rice A and Rosenstiel T 2016 Mechanisms of methane transport through Populus trichocarpa *Biogeosci. Discuss.* bg-2016–60
[22] Carmo J B D, Keller M, Dias J D, Camargo P B D and Crill P 2006 A source of methane from upland forests in the Brazilian Amazon *Geophys. Res. Lett.* **33** L04809

IOP Conf. Series: Earth and Environmental Science **138** (2018) 012003 doi:10.1088/1755-1315/138/1/012003

# Partitioning of net carbon dioxide flux measured by automatic transparent chamber

**EA Dyukarev**

Institute of Monitoring of Climatic and Ecological System of the Siberian Branch Russian Academy of Sciences, Akademicheskii 10/3, Tomsk, Russia

Email: egor@imces.ru

**Abstract**. Mathematical model was developed for describing carbon dioxide fluxes at open sedge-sphagnum fen during growing season. The model was calibrated using the results of observations from automatic transparent chamber and it allows us to estimate autotrophic, heterotrophic and ecosystem respiration fluxes, gross and net primary vegetation production, and the net carbon balance.

## 1. Introduction

The West Siberia peatlands are important components of the terrestrial carbon cycle and store about 70 Pg of carbon in peat soil [1, 2]. Although the size of the carbon reservoir is considerable, the role of peatlands in the global carbon budget has not been studied deep enough [3].

The gaseous exchange between the atmosphere and the peatlands is dominated by photosynthetic fixation of $CO_2$ from the atmosphere and by soil and vegetation respiration losses of $CO_2$. The balance between them is known as the net ecosystem exchange (NEE) of $CO_2$ [4, 5]. The other major gaseous loss of C into the atmosphere is $CH_4$ which is produced via anoxic decay of the soil organic matter [6, 7]. The loss of C into the fluvial system occurs via export of dissolved and particulate organic carbon, and dissolved gases ($CO_2$ and $CH_4$). Numbers of environmental factors play important roles in governing the rate of net $CO_2$ exchange in peatlands and the expected climate change may affect these regulating factors [4 -7].

The present climate warming in the Northern hemisphere [8, 9] may increase ecosystem carbon uptake by reducing cold-temperature constraints for plant carbon assimilation and growth. Soil warming also accelerates carbon losses due to exponential effect of temperature on soil respiration [10]. Possible changes in temperature and humidity may alter the peatland carbon budget significantly. Estimation of carbon exchange rates between peatlands and the atmosphere as well as the environmental controls on this exchange under existing climatic changes are important scientific objectives.

Modeling approaches are useful to separate the observed net ecosystem exchange into gross primary production (GPP) and total ecosystem respiration (ER) components, since it provides a better diagnostic of ecosystem processes and their regulating factors [8, 11, 12]. Carbon balance models are used to study the relative importance of different environmental factors, and for filling the gaps in time series to calculate daily and annual carbon budget [11, 12]. Partitioning of the NEE flux is also needed for better understanding of interannual and spatial variability of the carbon fluxes.

The purpose of this study is to develop a mathematical model for estimating $CO_2$ exchange fluxes of a meso-oligotrophic fen at West Siberia, and to calibrate the model using field data on $CO_2$ fluxes obtained using automatic transparent chamber and infra-red gas analyzer.

## 2. Methods

The measurement site is located in the south of West Siberia (Russia) at the Bakcharskoe bog (area 1400 $km^2$) in the interfluves of the rivers Iksa and Bakchar in Bakcharsky district of Tomsk region, Russia [13]. The observation site was arranged at mesooligotrophic open sedge-sphagnum fen with continuous moss cover of *Sphagnum magellanicum.* Herbs and sedges cover about 50% of the area and they are represented mostly by *Eriophorum vaginatum, Carex rostrata, Carex limosa* species [2].

An automated soil $CO_2$ flux system LI-8100A (Li-Cor Biogeoscience, USA) with transparent long-term chamber LI-8100-104 has been used for carbon dioxide emission measurements during field campaigns in 2014-2015. Measurements were conducted two times in a month during daytime only. Intense field campaigns were held in August 15-17 and August 21-25 for a diurnal observation of $CO_2$ fluxes.

The chamber was installed at plastic cylindrical basement deepened into the peat to 15 cm. The vegetation cover under the chamber consists of mosses and sedges only. The carbon dioxide fluxes were estimated by increase of $CO_2$ concentration in the chamber during an exposition. An exposition was 5 min, but only two first minutes were used for total flux calculation when $CO_2$ concentration changes according to linear manner. The measurements were automatically repeated every 20 minutes. Simultaneous observations of air, surface and peat temperatures, incoming photosynthetically active solar radiation (PAR), atmospheric pressure, air water content, water table level and precipitation were made.

The measured NEE was partitioned into GPP and ER fluxes:

$$NEE = ER - GPP.$$

The total ecosystem respiration was subdivided into heterotrophic respiration (HR) and autotrophic respiration (AR). The plant parts and microbes have different temperature sensitivities and temporal decoupling of autotrophic and microbial processes might therefore enhance the model of the temperature response of respiration. Each respiration component was modeled using an exponential equation (2-3) widely used for ER variations explanation [12, 14 - 16]

$$ER = HR + AR; \quad (1)$$

$$HR = E_H \times exp(k_H \times T_a); \quad (2)$$

$$AR = LAI \times E_A \times exp(k_A \times T_a), \quad (3)$$

where $T_a$ is air temperature ($^oC$), $E_H$ and $E_A$ are basal levels of heterotrophic and autotrophic respiration at 0 $^oC$, respectively, $k_H$ and $k_A$ are temperature sensitivity coefficients. Air temperature was used as an explanatory factor. Autotrophic respiration started to increase proportionally to LAI with the growth of green vegetation.

A rectangular hyperbolic function was used for the light response of GPP in daytime [12, 14 - 15]:

$$GPP(PAR) = LAI \times \alpha \times PAR \times G_m / (\alpha \times PAR + G_m), \quad (4)$$

where $\alpha$ is the initial slope of the light response curve at low light (photosynthetic efficiency), $G_m$ is the theoretical maximum rate of photosynthesis at infinite PAR (photosynthetic capacity).

IOP Conf. Series: Earth and Environmental Science **138** (2018) 012003 doi:10.1088/1755-1315/138/1/012003

In this study NEE is negative when the value of GPP exceeds the ER value and there is a net removal of carbon dioxide from the atmosphere. NEE is positive when the ER value exceeds the GPP value and the carbon dioxide is released from the ecosystem into the atmosphere.

## 3. Results and discussion

The mathematical model was calibrated using field observation data. The multiply optimization procedure was performed in MATLAB software using *fminsearch* function. The minimum of the unconstrained multivariable function was found using the derivative-free method. The root-mean-square error was used as minimizing function. The following model parameters were estimated: $G_m$ = 1603.7 $mgCO_2\ m^{-2}\ h^{-1}$, $\alpha$ = 0.71 $mgCO_2\ m^{-2}\ h^{-1}$/ $\mu mol\ m^{-2}\ s^{-1}$, $E_H$ = 59.3 $mgCO_2\ m^{-2}\ h^{-1}$, $E_A$ = 39.5 $mgCO_2\ m^{-2}\ h^{-1}$, $k_H$ = 0.047 $^oC^{-1}$, $k_A$ = 0.024 $^oC^{-1}$. The temperature sensitivity coefficient for heterotrophic respiration is two times greater than the corresponding coefficient for autotrophic respiration. The basal level of heterotrophic respiration exceeds the basal respiration of plants by 67%. There is a good agreement between the model output and the observations in daytime but it is not so satisfactory for night period. The observed night time variations of ecosystem respiration are much greater than the modeled ones.

The proposed mathematical model may be used to calculate daily and annual carbon dioxide fluxes for the studied ecosystem. For the variations of the modeled net daily $CO_2$ fluxes in 2016 see Figure 1. The NEE values increase from 0.64 gC $m^{-2}\ d^{-1}$ in the beginning of April to 2.37 gC $m^{-2}\ d^{-1}$ in the middle of July. The share of plant respiration in ER was accounted as 0.5 in the mid-summer and less than 0.3 at the beginning and the end of the vegetation season. The maximal GPP (5.02 gC $m^{-2}\ d^{-1}$) in 2016 was obtained in June 23, the minimal GPP (0.39 gC $m^{-2}\ d^{-1}$) – at the beginning of April. The simulated NEE varies significantly from 0.89 to -2.86 gC $m^{-2}\ d^{-1}$ during a growing season due to variations of the main controls. The average NEE value is -0.81 gC $m^{-2}\ d^{-1}$. Therefore, the carbon assimilation by peatland vegetation is greater than the total ecosystem respiration during a vegetative season. The net carbon accumulation in 2016 for the studied ecosystem was estimated as 149 gC $m^{-2}\ d^{-1}$. The net primary production (NPP = GPP - AR) was 300 gC $m^{-2}\ d^{-1}$.

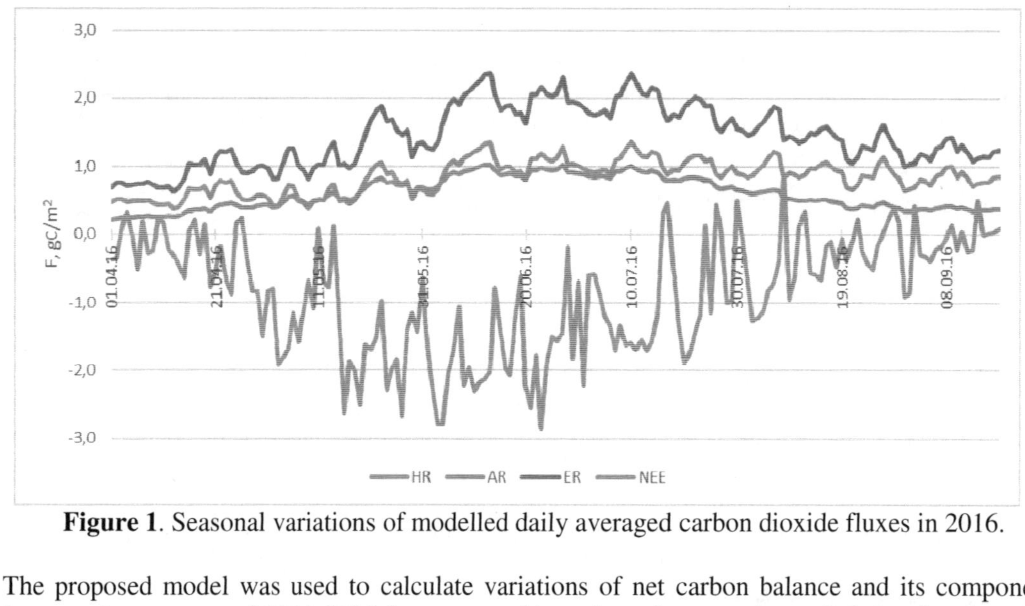


**Figure 1**. Seasonal variations of modelled daily averaged carbon dioxide fluxes in 2016.

The proposed model was used to calculate variations of net carbon balance and its components during growing seasons of 2011-2016 for mesotrophic sedge-sphagnum fen at Bakcharskoe bog. The data on air temperature and incoming solar radiation were taken from the nearest weather station. It was found that the studied ecosystem act as a stable sink for the atmospheric carbon (Table 1) accumulating annually from 112 to 148 gC $m^{-2}$.

The release of carbon by heterotrophic and plants respiration according to simulation was estimated at 146.2 and 105.6 gC $m^{-2}$ respectively. The total ecosystem respiration was evaluated as 251.9 gC $m^{-2}$. The gross primary production of fen vegetation (386.6 gC $m^{-2}$) exceeds the ecosystem respiration. The net primary production values (281 gC $m^{-2}$) are in good agreement with the data obtained by the vegetation clipping methods. The net carbon accumulation was estimated as 134.7 gC $m^{-2}$ annually (see Table 1).

**Table 1.** Modelled annual net carbon balance (gC $m^{-2}$) and its components for mesotrophic sedge-sphagnum fen at Bakcharskoe bog in 2011-2016.

| Year | GPP | AR | HR | ER | NPP | NEE |
|---|---|---|---|---|---|---|
| 2011 | 379,4 | 110,2 | 157,1 | 267,3 | 269,1 | -112,1 |
| 2012 | 392 | 108 | 151,5 | 259,5 | 284 | -132,6 |
| 2013 | 367,3 | 99,4 | 134,1 | 233,4 | 268 | -133,9 |
| 2014 | 387,3 | 103,1 | 139,3 | 242,4 | 284,1 | -144,8 |
| 2015 | 386,3 | 105,6 | 144,6 | 250,2 | 280,7 | -136 |
| 2016 | 407,4 | 107,5 | 151,3 | 258,8 | 300 | -148,7 |
| Mean | 386,6 | 105,6 | 146,3 | 251,9 | 281,0 | -134,7 |

## 4. Conclusion

A model of net ecosystem exchange was performed to study the influence of different environmental factors and to calculate daily and growing season carbon budget. The model uses air temperature, incoming photosynthetically active radiation and leaf area index as the explanatory factors for gross primary production, heterotrophic and autotrophic respiration. The model coefficients were calibrated using data collected by automated soil $CO_2$ flux system with clear long-term chamber at mesotrophic open sedge-sphagnum fen. The results allow us to calculate different temperature sensitivity of heterotrophic and plant respiration. It indicates that the proposed model is a promising tool for a better understanding of ecosystem biogeochemical processes. The model may be applied to the simulation of intra-annual variations in $CO_2$ flux components in peatland ecosystems.

**Acknowledgments**
This study was supported by Russian Foundation for Basic Researches (grant numbers 16-07-01205 and 16-45-700562) and Russian Academy of Sciences (project number VIII.77.1.1).

**References**

[1] Sheng Y, Smith L C, MacDonald G M, Kremenetski K V, Frey K E , Velichko A A, Lee M, Beilman D W and Dubinin P 2004 *Glob. Biogeochem. Cycles* **18** GB3004
[2] Golovatskaya E A, Dyukarev E A, Ippolitov I I and Kabanov M V 2008 *Rep. of the Russian Acad. of Sci.* **418(1)** 187–90
[3] Alekseeva M N, Preis Y I and Dyukarev E A 2015 *Bull. of the Tomsk Polytechnic Univer., Geo Assets Engineering* **32(4)** 81–90
[4] Bubier J L, Crill P M, Mosedale A, Frolking S and Linder E 2003 *Glob. Biogeochem. Cycles.* **17(2)** 1066
[5] Golovatskaya E A and Dyukarev E A 2012 *Eurasian Soil Sci. J.* **45(6)** 588–97 doi:10.1134/S106422931206004X
[6] Saunois M., Bousquet P., Poulter B. et al. 2016 *Earth Syst. Sci. Data* **8** 697–751

IOP Conf. Series: Earth and Environmental Science **138** (2018) 012003 doi:10.1088/1755-1315/138/1/012003

doi:10.5194/essd-8-697-2016
[7] Veretennikova E E, Dyukarev E A 2017 *Russian Meteorol. and Hydrol.* **42(5)** 319–26 doi:10.3103/S1068373917050077
[8] Ciais P, Sabine C, Bala G. et al. 2013 *Carbon and Other Biogeochemical Cycles. In: Climate Change 2013: The Physical Science Basis. Contribution of Working Group I to the Fifth Assessment Report of the Intergovernmental Panel on Climate Change* (Cambridge University Press, Cambridge, United Kingdom and New York, NY, USA) p 1535
[9] Dyukarev E A 2015 *Earth's Cryosphere* **19(3)** 45–51
[10] Golovatskaya E A and Dyukarev E A 2011 *Russian Meteorol. and Hydrol.* **36(6)**
[11] Falge E, Baldocchi D, Olson R, et al. 2001 *Agric. For. Meteorol.* **107** 43–69
[12] Dyukarev E A 2017 *Agric. For. Meteorol.* **239** 236–48 doi:10.1016/j.agrformet.2017.03.011
[13] Dyukarev E A, Golovatskaya E A, Duchkov A D and Kazantsev S A 2009 *Russian Geol. and Geoph.* **50(6)** 579–86
[14] Mäkelä A, Hari P, Berninger F et al. 2004 *Tree Physiol.* **24(4)** 369–76
[15] Laine A, Riutta T, Juutinen S, et al. 2009 *Ecol. Modell.* **220** 2646–55 doi:10.1016/j.ecolmodel.2009.06.047
[16] Kandel T P, Elsgaard L, Larke P E 2013 *Bioenergy* **5(5)** 548–61 doi:10.1111/gcbb.1202

IOP Conf. Series: Earth and Environmental Science **138** (2018) 012004 doi:10.1088/1755-1315/138/1/012004

# Short-term standard litter decomposition across three different ecosystems in middle taiga zone of West Siberia

**Nina V Filippova*[1] , Mikhail V Glagolev[2]**

UNESCO chair Environmental dynamic and global climate change, Yugra State University, Chekhova street, 16, Khanty-Mansiysk, 628012, Russia
Faculty of soil science, Moscow State University, Leninskie Gory 1-12, 119991, Russia

*Email: n_filippova@ugrasu.ru

**Abstract**. The method of standard litter (tea) decomposition was implemented to compare decomposition rate constants (*k*) between different peatland ecosystems and coniferous forests in the middle taiga zone of West Siberia (near Khanty-Mansiysk). The standard protocol of TeaComposition initiative was used to make the data usable for comparisons among different sites and zonobiomes worldwide. This article sums up the results of short-term decomposition (3 months) on the local scale. The values of decomposition rate constants differed significantly between three ecosystem types: it was higher in forest compared to bogs, and treed bogs had lower decomposition constant compared to Sphagnum lawns. In general, the decomposition rate constants were close to ones reported earlier for similar climatic conditions and habitats.

## 1. Introduction

### *1.1. Litter decomposition in ecosystems.*

Litter decomposition determines the nutrient cycles in ecosystems and represents one of the largest fluxes in the global terrestrial carbon cycle [1]. In this context, the peatland ecosystems deserve special attention due to significant carbon storage and their real response to the changing environment.

In the recent past several litter decomposition studies using native litters were performed in the diverse peatland ecosystems in West Siberia [2, 3, 4, 5]. The basic method of litter bags with native substrates was implemented in these studies. About 30 litter types were studied in total (including dwarfshrubs, herbs, mosses and lichens; their above ground and belowground parts).

The decomposition rates of different litters were obtained along with evaluation of environmental parameters, such as temperature, micro-habitat conditions, elemental composition of the substrates and soils, depth below the surface and others. The dynamics of decomposition rates through the course of time were described for some of litter types.

Despite the importance of using native litter types in decomposition experiments, this approach does not allow to make comparisons across the sites and hence hamper the identification of the key drivers of litter decomposition on a global scale. A combined approach of local and standard litter was implemented in several large decomposition studies such as DECO and LIDET. In order to prove the possibility of using this technique, M. Didion with co-authors [6] compared long-term decomposition rates of local litter and tea (Lipton Rooibos and Green tea) in the Austrian Alps and found a similar responses in decomposition to the changing climatic conditions and associated shift in the vegetation.

In this context, recently started Global Litter Decomposition Study, TeaComposition, within the International Long Term Ecological Research (ILTER) aims to study the long-term litter C dynamics and its key drivers at the present and predicted climate scenarios worldwide by using standard substrate (tea).

*1.2. Purpose of the study and hypotheses.*
In the current study we aim to investigate the standard substrate (tea) to compare decomposition rate constants (k) between different peatland ecosystems and coniferous forests in the middle taiga zone of West Siberia. We hypothesize that 1) decomposition rate constant would be higher in coniferous forests compared to peatlands (due to better aeration conditions and better nutrient content of the soil), 2) within peatland ecosystems the waterlogged lawns would have lower decomposition constant compared to treed bogs, referring to series of experiments previously made under the same conditions.

*1.3. Abbreviations.*
MFS – Mukhrino Field Station; FS – Field Station; *k* – Decomposition constant; S – Stabilization factor [7]; $a_r$ – Decomposable fraction of rooibos tea [7].

## 2. Materials and methods

*2.1. Site description.*
The experiment on tea decomposition was held in the vicinities of Mukhrino field station of Yugra State University [8] which is located in the middle taiga zone of West Siberia on the left terrace of the Irtysh River in 30 km South-West from Khanty-Mansiysk (60.8890°N; 68.7027°E). In the studied area, the vegetation is mainly characterized by coniferous forests and raised *Sphagnum* bogs. The forested areas develop on well-drained locations whereas large areas in the station vicinities are occupied by peatland landscapes. The oligotrophic peatlands are the most prevalent and represented by treed pine-dwarf shrubs-*Sphagnum* communities and graminoid-*Sphagnum* lawns. The water table may fluctuate between 0-15 cm depending on season and precipitation rate at the lawns, while the treed bogs have thicker acrotelm layer where the water level is 30-50 cm below the surface. The soil (*Sphagnum* peat) has nearly equal characteristics in the treed bogs and the *Sphagnum* lawns with low pH values (pH 3-4), high percentage of organic carbon (49-58%) and low ash content (1.2-2.6 %). The soil nutrients, calculated for the upper layer of peat in this bog were: K – 224, Na – 177, Ca – 1347, Mg – 278, P – 206, Fe – 791 mg/kg of dry weight (all soil characteristics are according to [20]).
The forests are made up of several coniferous trees and the undergrowth of various small herbs, ericoid shrubs and feather mosses. The coniferous forests are replaced by their secondary communities from *Betula* spp. and *Populus tremula*. The underlying soils are spodosols (= podzolic soils) formed on clayey loam parent material and are characterized by thin organic layer and low pH values [9].

*2.2. Litter bag experiment.*
The general scheme of TeaComposition project was implemented in the study [7]. Two plots (200 $m^2$) (waterlogged graminoid-Sphagnum lawn and relative dry pine-dwarf shrubs-Sphagnum treed bog) differing mainly in water regime and associated vegetation composition were located within the raised bog (figure 1). The third plot was located in well-drained coniferous-deciduous mixed forest (see table A1 in Appendix for details of vegetation, plots coordinates and soil properties). Lipton tetrahedron-shaped synthetic tea bags containing about 2 g of green or rooibos tea were prepared for incubation according to protocol described in [7]. At each site two replicas were established with 32 tea bags from each type buried inside a square 4 $m^2$ for incubation over the period of 3 years with the retrieval time after 3, 12, 24, and 36 months. Before the incubation, the tea was dried at 70°C for 48 hours and the initial weight was recorded. The tea was incubated in the aerated acrotelm layer (0-5 cm below the Sphagnum surface) at the two bog sites and under the litter layer (0-5 cm, approximately in organic

layer as it is poorly distinguished in the podzolic soils) at the forest site in June 2016. After 3 months of incubation the first 8 bags from each plot were collected, cleaned from earth and roots, dried and weighed.

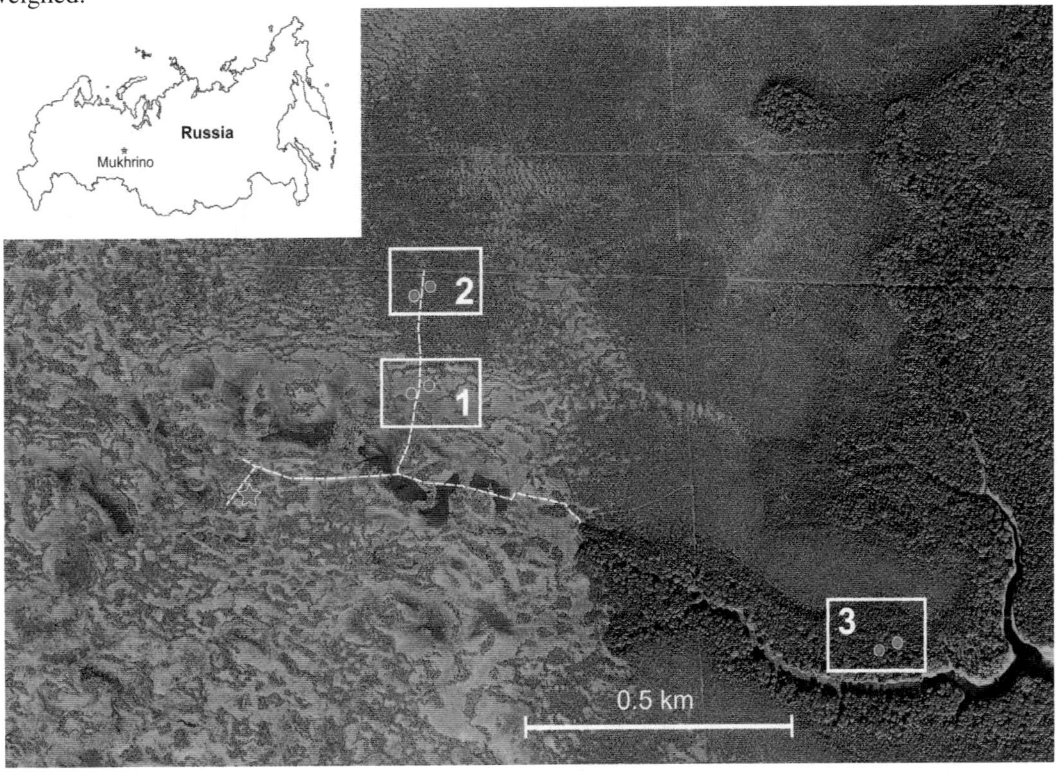


**Figure 1.** Location of plots and replicas in the vicinities of Mukhrino FS: Plot 1 – *Sphagnum* lawn of raised bog, Plot 2 – treed community of raised bog, Plot 3 – forest; the star marks the position of a micro-meteorological station MFS; the dotted line – a boardwalk through the experimental polygon of MFS in the bog

### *2.3. General climate characteristics and monitoring of climatic parameters.*

The general climate characteristic in the Khanty-Mansiysk vicinities is continental subarctic or boreal with the average annual temperature being –1.3°C, the mean temperature of the coldest month (January) –19.8°C, the warmest month is July with its average of 18°C. The mean annual precipitation is 553 mm [10].

The weather conditions were measured by automatic micro-meteorological station of Mukhrino FS located in the bog, including Rotrinic Hygroclip S3 for measuring air temperature and HOBO rain gauge RG3-M for measuring liquid precipitation. The mean temperatures of May-September were higher by 2-4°C compared to the values of the average climate in Khanty-Mansiysk. The number of rains was 20-50% lower in summer (May-August) but September had 30% more rains compared to the average climate characteristics.

### *2.4. Data analysis.*

The calculation of *k* (decomposition constant), S (stabilization factor) and $a_r$ (decomposable fraction of rooibos or green tea) was made as described in [7]. The values of differences between three plots were calculated by Wilcoxon rank sum test [12] (this test is equivalent to the Mann-Whitney U test) using RANKSUM MATLAB-function.

IOP Conf. Series: Earth and Environmental Science **138** (2018) 012004 doi:10.1088/1755-1315/138/1/012004

## 3. Results

### *3.1. Visual characteristic of the decomposed tea.*

The visual characteristics of the bags content differed significantly among the plots as it was seen by direct observation and using lens (table B1 in the Appendix). Generally, green tea was more decomposed, it had darker color and was more homogenized compared to rooibos. The tea bags in lawns had almost intact structure of soaked tea, no roots or fungal mycelium were observed (except for two bags in Lawn 1 and Lawn 2, which were buried on a small hummock under more aerated conditions). The bags in the treed bog had darker color, green tea was very dark, homogenous and slimy, all bags with green tea were covered abundantly by some species of Myxomycota (dark slimy mass on the surface and abundant spores and capillitium threads were seen under the microscope) and rooibos was in several cases covered abundantly by *Aspergillus* sp., no or minor roots were observed. The tea in forest was similar to treed bog but green tea was more intact compared to treed bog, abundant fungal mycelium and mycorrhizal roots were observed on the surface and inside the bags (roots were carefully removed).

### *3.2. Litter mass loss.*

The mass loss after 3-months incubation of green tea was significantly higher (average 72%) compared to rooibos (average 27%) (figure 2A). With regard to the effect of the ecosystem type, there were significant differences between all three ecosystem types for green tea, and between forest/treed bog and lawn/treed bog for rooibos tea (table 1). The probability of overlap between median values of mass loss for forest and lawn was high (24%) and statistically insignificant. It is important to note that the mass loss of green and rooibos varied differently between the ecosystem types (e.g. mass loss for green tea decreased in a sequence lawn>treed bog>forest while for rooibos tea in a sequence lawn~forest>treed bog).

### *3.3. Stabilization factor and decomposition rate constant.*

The stabilization factor and decomposition rate differed significantly between all three ecosystem types (figure 2B, table 2, table C1 in the Appendix). The highest probability of overlap between median values was only 0.3%. The largest stabilization was in forest, intermediate in treed bog and the lowest in the lawn. The decomposition rate constant was the largest in forest and the lowest in the treed bog.

**Table 1.** Wilcoxon rank sum test for mass loss of two tea types in different ecosystems.

| Ecosystem type | Probability of overlap between median values | |
|---|---|---|
| | **Green tea** | **Rooibos tea** |
| Forest/Lawn | 0.0002 | 0.2370 |
| Forest/Treed bog | 0.0002 | 0.0002 |
| Lawn/Treed bog | 0.0019 | 0.0002 |

**Table 2.** Wilcoxon rank sum test for *S* and *k* values in different ecosystems.

| Ecosystem type | Probability of overlap between median values | |
|---|---|---|
| | **S** | ***k*** |
| Lawn/Treed bog | 0.003 | 0.0006 |
| Lawn/Forest | 0.0002 | 0.0003 |
| Forest/Treed bog | 0.0002 | 0.0002 |

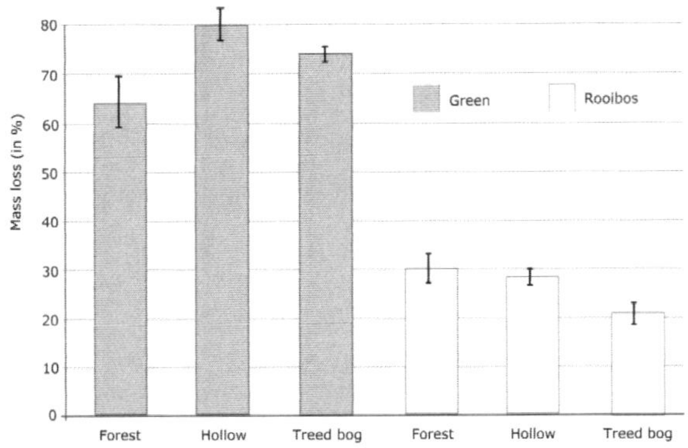


**Figure 2A.** Mean numbers of mass loss (in %) after 3 months of decomposition: two types of tea bags (rooibos, green) in three habitats (treed bog and hollow of raised bog and coniferous forest), black bars show standard deviations.

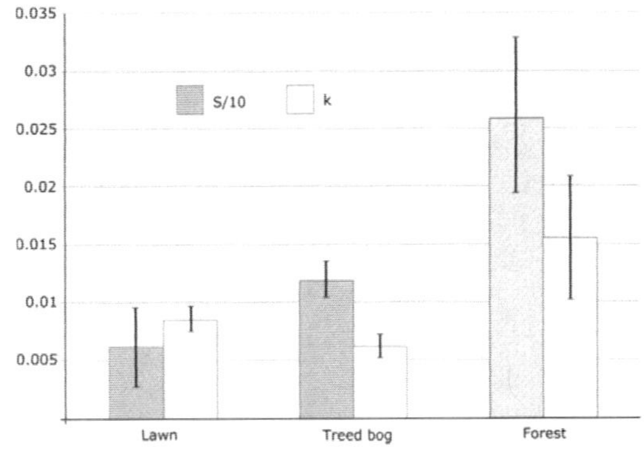


**Figure 2B.** Mean numbers of stabilization factor (S/10) and decomposition rate constant ($k$, day$^{-1}$) for three studied ecosystem types, black bars show standard deviations.

## 4. Discussions

According to general view, accumulation of peat is provided by an imbalance between slow rates of decomposition and rate of net primary production in peatlands. This is determined however by multiple factors, e.g. poor nutrient content, cool and anoxic conditions, high refractory content of peatland litters, and others [12, 13]. We tested the hypothesis of slower decomposition rates in bog compared to forest when the influence of litter content is reduced by the standard litter (tea).

We hypothesized that decomposition rate constant would be higher in coniferous forests compared to peatlands. This assumption was confirmed by the calculations of decomposition rate constant which was almost twice higher in forest compared to both bog sites. The higher decomposition rate in forest compared to peatland could be explained by better aeration conditions and better nutrient content of the soil and is consistent with the general rule of slow decomposition rates in peatlands. However, the forest site had also the highest stabilization factor, which is probably explained by unusually dry summer, which preserved high fraction of green tea (but not rooibos tea!) in forest. It is likely that the local weather conditions could significantly influence the short-term decomposition rates. We assume therefore that the experiment should be repeated in the following years to verify the decomposition rates in relation to weather conditions.

The literature review shows different estimates when comparing the decomposition rates between upland forest and peatlands. T. Moore and N. Basiliko [12] in a study of three tree leaf types in a temperate swamp and adjacent deciduous forest in Canada showed higher decomposition rate in forests in only half of the plots and the reverse in the other half. T. Moore with coauthors [14] made litter transplant experiment (using the same litters, e.g. 10 foliar litter types and wood, under different drainage classes) and showed no significant difference between decomposition rates in uplands and peatlands. However, they report significant differences for two litter types decomposing faster either in upland or peatland conditions. The authors suggest that longer term (>6 years) differences in decomposition rate (when the litter would become buried under anoxic conditions) and differences in litter quality account for larger accumulation of organic matter in peatlands compared to forest.

As we mentioned above, the mass loss sequences in three ecosystems varied for green and rooibos tea. The use of two litter types for estimation of $k$ at a single point in time was based on assumption that both tea types will equally react to the environmental parameters [7]. As in our study this condition was not satisfied, it could cause incorrect calculation of $a_r$ based on stabilization factor. However, these differences on short-time scale could be caused by insufficient decomposition time for slowly decomposable rooibos tea. The following retrieval of the next sequences of bags (after 1, 2 and 3 years of decomposition) could solve this problem.

Regarding the differences between the peatland habitats, our hypothesis was that the waterlogged lawns would have lower decomposition constant compared to treed bog. The calculation of decomposition constant showed the opposite: the lawns had higher decomposition rate compared to treed bogs. The stabilization factor corresponded to the decomposition rate: it was low in lawns and higher in treed bogs. According to the series of experiments in similar habitats in taiga zone of West Siberia the decomposition in lawns is slower compared to treed bogs due to high water level and consequently anaerobic conditions. This was shown in litter transplant experiments for several plant litters incubated in both habitats. N. Kosykh with coauthors [2] compared decomposition rates of a dwarfshrub leaves (*Chamaedaphne calyculata*) on hummocks and in lawns and showed higher decomposition rate in better aerated conditions of hummocks. The mass loss of peat samples was significantly higher on hummocks compared to hollows in peatlands in different zones of West Siberia [15]. The unusually dry summer could influence the opposite results in our case: the lawns were drier during this year which created thicker aerated layer and higher decomposition rates (and probably treed bogs were at the same time too dry for normal decomposition rate). Possible relationships between the decomposition rate and the water level in lawns and hummocks will be determined after three years of decomposition experiment using data on water level collected by Mini-Diver sensors installed in close proximity to the experimental plots.

There are many theoretical models to estimate the decomposition rate constant depending on humidity, for example [16, 17, 18, 19]. A simple relationship developed by A. Van der Linden with co-authors [16] shows, that decomposition rate constant is about 3 times higher under optimal conditions of humidity compared to the conditions of full moisture saturation. Considering the forest conditions being optimal, and the bog (lawn and treed bog) being fully saturated, the $k$ values estimated in our experiment are consistent with this model.

The values of the decomposition rate constant ($k$) resulting from our work were compared with the values for boreal zone from literature. The mean values of $k$ in forest, lawn and treed bog were 0.015, 0.008 and 0.006 $day^{-1}$ respectively. T. Moore and N. Basiliko [12] report $k$ values in peatlands (calculated on a 6 year basis) from 0.000027[3] $day^{-1}$ (spruce needless) to 0.0012 $day^{-1}$ (birch leaves) when litter decays rapidly. The values of slowly decomposing litters (*Sphagnum fuscum*, wood blocks) could range from 0.00 to 0.000027 $day^{-1}$ and 0.000027 to 0.000082[4] $day^{-1}$ respectively [12: table 7.1]. N. Kosykh with co-authors in a study performed in close vicinities to our experimental site recorded similar values of decomposition constant (based on two months of decomposition) for leaves of *C. calyculata* (k=0.0067 $day^{-1}$)[5], but higher value for roots of the same plant (k=0.009 $day^{-1}$) [2]. The values of $k$ and S estimated in our study correspond to similar ecosystem types measured by J. Keuskamp with co-authors [7: figure 3] (such as peatlands sites in Ireland and in the Netherlands, calculations based on 60-90 days of decomposition).

The significant difference (for 1-3 orders of magnitude) between our decomposition constants and reported by T. Moore and N. Basiliko [12] could be explained by difference in decomposition time (3 months vs. 6 years). The decomposition process is described by formula (1) and includes two stages of decomposition of labile and recalcitrant compounds:

$$W(t)=ae^{-k_1 t} + (1-a)e^{-k_2 t}$$

where $W(t)$ is the weight of the substrate after incubation time $t$, $a$ is the labile, $(1-a)$ is the recalcitrant fraction of the litter, $k_1$ and $k_2$ are decomposition rate constants of the labile and recalcitrant fractions accordingly [7]. During the short-time experiments (90 days in our case) $k_2$ is very low and cannot be estimated. On the other hand, in long-lasting experiments (6 years) where the process is approximated by simple exponential decay model [12] the decomposition constant will either be equal $k_2$ (and

[3] The original values were in another dimension: 0.01 (spruce needless) and 0.45 (birch leaves) $year^{-1}$.

[4] The original values were 0.00 to 0.01 and 0.01 to 0.03 $year^{-1}$.

[5] The original values were: *C. calyculata* ($k$=0.08) and roots ($k$=0.11) $month^{-1}$.

therefore very low) or intermediate between $k_1$ and $k_2$. In contrast, the values of $k$ estimated during the first months of decomposition [2] are very close to our 90-days estimates.

The mass loss of tea bags in our study was higher compared to earlier experiments in the Northern Limestone Alps of Austria in forest and grassland communities (subalpine to alpine climate zones) [6]. M. Didion with coauthors report the mean values being 30% for green tea and 7% for rooibos tea for a year of decomposition, while we measured mean values 72% for green tea and 28% for rooibos tea for 3 month of decomposition. We cannot explain these differences, considering seemingly harsher conditions in our region (lower mean annual temperatures and precipitation values). However, similar mass losses of approximately 70% for green tea and 30% for rooibos tea were also reported by J. Keuskamp with co-authors [8] in experimental conditions after 140 days of decomposition. The mean temperature of incubation period in our study was comparable to that of experimental conditions (15°C) without taking into account daily fluctuations in natural conditions.

The mass loss of standard litter (tea) was compared with estimates of native litters mass loss on a located nearby peatland [2]. The mass losses of *Sphagnum* were 7-15%, woody parts of dwarfshrubs – 18%, rhizomes of sedges – 15-30%, roots of dwarfshrubs – 20-40%, roots of sedges and herbs – 30-40%, green parts of herbs and dwarfshrubs – 50-86%, and rhizomes of *Menyanthes trifoliata* – 80% of their initial weight after 1 year of decomposition. Comparing to this native substrates, the mass loss of rooibos tea was close to slowly decomposing substrates such as wood and *Sphagnum*, and green tea mass loss corresponded to easily decomposable green parts of plants and rhizomes.

## 5. Conclusions

We estimated mass loss, decomposition rate constant and stabilization factor in three major ecosystem types in middle taiga zone of West Siberia using standard litter (tea) short-term decomposition experiment. The values of decomposition rate constants differed significantly between three ecosystem types (coniferous forest, treed *Sphagnum* bog and *Sphagnum* lawn). The decomposition rates in forest were higher compared to bog. Between bog sites, treed bog had lower decomposition rates compared to lawn, which could nonetheless be biased by unusually dry summer conditions. The experiment should be therefore repeated in future years to obtain unbiased data.

The mass loss and the decomposition rate constants were close to reported earlier for similar climatic conditions and habitats.

## Acknowledgements

We are thankful to Djukic Ika for providing contact with TeaComposition experiment and proofreading of the manuscript. Also great thanks for an anonymous reviewer for important remarks during the article's revision.

## Appendices

**Table A1.** Description of vegetation and soil properties of the experimental plots.

| Plot / Replica | Coordinates | Vegetation | Dominant plants with % cover (only species with > 5 % of cover shown) | Soil type and description | pH |
|---|---|---|---|---|---|
| 1/1 | 60.89325 N 68.68311 E | Graminoid-*Sphagnum* lawn bog | *Andromeda polifolia* (5), *Oxycoccus palustris* (3), *Carex limosa* (5), *Scheuchzeria palustris* (2), *Eriophorum russeolum* (3), *Sphagnum balticum* (30), *S. jensenii* (50), *S. papillosum* (10), *S. lindbergii* (5) | Sphagnum peat: 0-5 - aerated layer, penetrated by roots (could become anaerobic in spring and after heavy rainfall); 5-400 cm - anaerobic peat layer. | 3-4 |
| 1/2 | 60.89307 N 68.68301 E | | | | |

| | | | | | |
|---|---|---|---|---|---|
| **2/1** | 60.89497 N 68.68335 E | Pine-dwarfshrubs-*Sphagnum* treed bog | *Pinus sibirica* (3), *P. sylvestris* (40), *Chamaedaphne calyculata* (10), *Ledum palustre* (30), *Vaccinum uliginosum* (6), *Oxycoccus microcarpa* (5), *Rubus chamaemorus* (20), *Sphagnum fuscum* (95), *S. angustifolium* (5) | Sphagnum peat: 0-30 (50) cm - aerated layer, penetrated by root; 50-400 cm - anaerobic peat layer | 3-4 |
| **2/2** | 60.89507 N 68.68339 E | | | | |
| **3/1** | 60.88911 N 68.69941 E | Coniferous-deciduous mixed forest | *Populus tremula* (50), *Pinus sibirica* (20*)*, *Picea obovata* (10), *Rubus saxatilis* (15), *Vaccinium vitis-idaea* (12), *Gymnocarpium dryopteris* (10), *Equisetum sylvaticum* (7), *Linnaea borealis* (6), *Hylocomium splendens* (15), *Pleurozium schreberi* (10) | Spodosols: $A_0$ - forest litter; 0-20 cm (A) clayey loam, penetrated by roots, light grayish color; 20-100 cm (B) clayey loam, dirty brown color; 100-200 cm - silt (loess), grayish, gleying. | 4-5 |
| **3/2** | 60.88897 N 68.69887 E | | | | |

**Table B1.** Initial and final weight of tea and description of its characteristics, presence of roots, and fungal mycelium after 3 months of decomposition.

| Bag # | Plot | Tea type | Initial weight | Final weight after 3 month | Mass loss % | Observations on tea quality, roots and fungal mycelium[a] |
|---|---|---|---|---|---|---|
| 1 | Lawn | green | 1.7158 | 0.4051 | 76 | Dark homogenous; weak roots; abundant Myxomycota |
| 2 | Lawn | green | 1.7708 | 0.4689 | 74 | Dark homogenous; weak roots; abundant Myxomycota |
| 3 | Lawn | green | 1.7064 | 0.3121 | 82 | Green slimy leaves; no roots; no mycelium |
| 4 | Lawn | green | 1.7107 | 0.3251 | 19 | Green slimy leaves; no roots; no mycelium |
| 17 | Lawn | rooibos | 1.8282 | 1.3160 | 28 | Sodden tea; no roots; no mycelium |
| 18 | Lawn | rooibos | 1.8116 | 1.2552 | 31 | Sodden tea; no roots; no mycelium |
| 19 | Lawn | rooibos | 1.7637 | 1.3008 | 26 | Sodden tea; no roots; no mycelium |
| 20 | Lawn | rooibos | 1.7577 | 1.2761 | 27 | Sodden tea; no roots; no mycelium |
| 33 | Lawn | green | 1.7364 | 0.3938 | 77 | Green slimy leaves; no roots; no mycelium |
| 34 | Lawn | green | 1.6893 | 0.3283 | 81 | Green slimy leaves; no roots; no mycelium |
| 35 | Lawn | green | 1.7478 | 0.3317 | 81 | Green slimy leaves; no roots; no mycelium |
| 36 | Lawn | green | 1.6676 | 0.3385 | 80 | Green slimy leaves; no roots; no mycelium |
| 49 | Lawn | rooibos | 1.8186 | 1.3091 | 28 | Sodden tea; no roots; no mycelium |
| 50 | Lawn | rooibos | 1.5515 | 1.0996 | 29 | Sodden tea; no roots; no mycelium |
| 51 | Lawn | rooibos | 1.7774 | 1.2511 | 30 | Sodden tea; no roots; no mycelium |
| 52 | Lawn | rooibos | 1.8384 | 1.3030 | 29 | Sodden tea; no roots; no mycelium |
| 65 | Treed bog | green | 1.7200 | 0.4700 | 73 | Dark homogenous; no roots; abundant Myxomycota |
| 66 | Treed bog | green | 1.7567 | 0.4365 | 75 | Dark homogenous; no roots; abundant Myxomycota |
| 67 | Treed bog | green | 1.7410 | 0.4832 | 72 | Dark homogenous; no roots; abundant Myxomycota |
| 68 | Treed bog | green | 1.7993 | 0.4599 | 74 | Dark homogenous; no roots; abundant Myxomycota |
| 81 | Treed bog | rooibos | 1.8376 | 1.4062 | 23 | Sodden tea; weak rooots; no mycelium |

| | | | | | | |
|---|---|---|---|---|---|---|
| 82 | Treed bog | rooibos | 1.7647 | 1.4489 | 18 | Sodden tea; weak rooots; no mycelium |
| 83 | Treed bog | rooibos | 1.7909 | 1.4155 | 21 | Sodden tea; weak rooots; no mycelium |
| 84 | Treed bog | rooibos | 1.8479 | 1.4857 | 20 | Sodden tea; weak rooots; no mycelium |
| 97 | Treed bog | green | 1.7336 | 0.4792 | 72 | Dark homogenous; no roots; abundant Myxomycota |
| 98 | Treed bog | green | 1.7735 | 0.4418 | 75 | Dark homogenous; no roots; abundant Myxomycota |
| 99 | Treed bog | green | 1.6357 | 0.4247 | 74 | Dark homogenous; no roots; abundant Myxomycota |
| 100 | Treed bog | green | 1.7667 | 0.4372 | 75 | Dark homogenous; no roots; abundant Myxomycota |
| 113 | Treed bog | rooibos | 1.8340 | 1.3974 | 24 | Sodden tea; some roots; *Aspergillum* |
| 114 | Treed bog | rooibos | 1.7567 | 1.3924 | 21 | Sodden tea; no roots; abundant *Aspergillum* |
| 115 | Treed bog | rooibos | 1.8244 | 1.4720 | 19 | Sodden tea; no roots; abundant *Aspergillum* |
| 116 | Treed bog | rooibos | 1.7303 | 1.3186 | 24 | Sodden tea; some roots; *Aspergillum* |
| 129 | Forest | green | 1.7243 | 0.6056 | 65 | Dark homogenous; many roots; abundant dark and white mycelium |
| 130 | Forest | green | 1.8772 | 0.6354 | 66 | Dark homogenous; many roots; abundant dark and white mycelium |
| 131 | Forest | green | 1.7754 | 0.8796 | 50 | Dark homogenous; some roots; abundant dark and white mycelium |
| 132 | Forest | green | 1.7810 | 0.5750 | 68 | Dark homogenous; some roots; abundant dark and white mycelium |
| 145 | Forest | rooibos | 1.8247 | 1.2767 | 30 | Sodden tea; many roots; mycelium and mycorrhiza |
| 146 | Forest | rooibos | 1.8152 | 1.2788 | 30 | Sodden tea; abundant roots; mycelium and mycorrhiza |
| 147 | Forest | rooibos | 1.8131 | 1.2637 | 30 | Sodden tea; many roots; some mycelium |
| 148 | Forest | rooibos | 1.8212 | 1.2041 | 34 | Sodden tea; many roots; some mycelium |
| 161 | Forest | green | 1.7035 | 0.6595 | 61 | Dark homogenous; some roots; Myxomycota |
| 162 | Forest | green | 1.7837 | 0.7595 | 57 | Dark homogenous; some roots; Myxomycota |
| 163 | Forest | green | 1.7735 | 0.6245 | 65 | Dark homogenous; some roots; Myxomycota |
| 164 | Forest | green | 1.6799 | 0.6130 | 64 | Dark homogenous; some roots; Myxomycota |
| 177 | Forest | rooibos | 1.8100 | 1.3286 | 27 | Sodden tea; some roots; some mycelium and mycorrhiza |
| 178 | Forest | rooibos | 1.7763 | 1.2510 | 30 | Sodden tea; some roots; some mycelium and mycorrhiza |
| 179 | Forest | rooibos | 1.8328 | 1.1993 | 35 | Sodden tea; some roots; some mycelium and mycorrhiza |
| 180 | Forest | rooibos | 1.7585 | 1.3202 | 25 | Sodden tea; some roots; some mycelium and mycorrhiza |

[a] Observations of fungal mycelium in mass made by naked eye, not attempt to estimate exact biomass of fungal mycelium was done in this study

**Table C1.** Values of stabilization factor (S) and decomposition rate ($k$) for retrieved samples after 3 months of decomposition.

| Plot | Replica | Vegetation Type | S | $k$ day$^{-1}$ |
|---|---|---|---|---|
| 1 | 1 | Lawn | 0.091068 | 0.008792 |
| 1 | 1 | Lawn | 0.124600 | 0.010776 |
| 1 | 1 | Lawn | 0.028199 | 0.007181 |
| 1 | 1 | Lawn | 0.037577 | 0.007758 |
| 1 | 2 | Lawn | 0.081147 | 0.008663 |
| 1 | 2 | Lawn | 0.042113 | 0.008606 |

| | | | | |
|---|---|---|---|---|
| 1 | 2 | Lawn | 0.036623 | 0.008643 |
| 1 | 2 | Lawn | 0.053024 | 0.008767 |
| 2 | 1 | Treed bog | 0.134633 | 0.007301 |
| 2 | 1 | Treed bog | 0.106768 | 0.004821 |
| 2 | 1 | Treed bog | 0.139254 | 0.006168 |
| 2 | 1 | Treed bog | 0.113912 | 0.005505 |
| 2 | 2 | Treed bog | 0.138426 | 0.007468 |
| 2 | 2 | Treed bog | 0.106388 | 0.005868 |
| 2 | 2 | Treed bog | 0.119012 | 0.005483 |
| 2 | 2 | Treed bog | 0.102855 | 0.007047 |
| 3 | 1 | Coniferous forest | 0.227974 | 0.013232 |
| 3 | 1 | Coniferous forest | 0.212233 | 0.012340 |
| 3 | 1 | Coniferous forest | 0.397756 | 0.026648 |
| 3 | 1 | Coniferous forest | 0.194095 | 0.015503 |
| 3 | 2 | Coniferous forest | 0.270107 | 0.011780 |
| 3 | 2 | Coniferous forest | 0.316074 | 0.016793 |
| 3 | 2 | Coniferous forest | 0.229166 | 0.018191 |
| 3 | 2 | Coniferous forest | 0.244673 | 0.009781 |

**References**

[1] Aerts R 1997 *Oikos* **79** 439–49

[2] Kosykh N P, Mironycheva-Tokareva N P and Parshina E K 2009 *Tomsk State Pedagogical University Bulletin* **3 (81)** 63–69

[3] Vishnyakova E K, Mironycheva-Tokareva N P and Koronatova N G 2012 *Tomsk State University Journal* **7 (122)** 87–93

[4] Mironycheva-Tokareva N P, Kosykh N P and Vishnyakova E K 2013 *Envyronmental dynamics and global climate change* **4-1**1–9

[5] Vishnyakova E K, Mironycheva-Tokareva N P 2014 *West Siberian peatlands and carbon cycle: past and present* 160–62

[6] Didion M, Repo A, Liski J, Forsius M, Bierbaumer M and Djukic I 2016 *Forests* **7 (167)** 10.3390

[7] Keuskamp J A, Dingemans B J J, Lehtinen T, Sarneel J M and Hefting M M 2013 *Methods Ecol Evol* **4** 1070–75

[8] Lapshina E D, Alekseyshik P, Dengel S, Filippova N V, Zarov E A, Filippov I V, Terentyeva I E, Sabrekov A F, Solomin Y R, Karpov D V and Mammarella I 2015 *Proc. of the 1st Pan - Eurasian Experiment (PEEX) Conf. and the 5th PEEX Meeting* **163** 236–40

[9] Bleuten W and Filippov I 2008 *Environmental dynamics and global climate change* **S1** 208–24

[10] Bulatov V I 2007 *Geography and ecology of Khanty-Mansiysk and its surroundings* (Khanty-Mansiysk: Izdatelstvo OAO "Infirmazionno-izdatelskiy zentr")

[11] Sharaf M A, Illman D L and Kowalski B R 1986 *Chemometrics* (New York: John Wiley & Sons)

[12] Moore T, Basiliko N 2006 Decomposition in Boreal Peatlands *Boreal peatland ecosystems* ed R K Wieder, P H Vitt (Berlin: Springer) 125–43

[13] Chernov I Ju 2013 *Microbial communities functioning in oligotrophic peatlands – the analysis of the causes of slow peat decomposition* (Moscow: Tovarishhestvo nauchnyh izdanij KMK)

[14] Moore T R, Trofymow J A, Siltanen M, Prescott C and Group C W 2005 *Can. J. For. Res.* **35** 133–42

[15] Koronatova N G and Milyaeva E V 2013 *Interexpo Geo-Siberia* **2-4** 166–71

[16] Van der Linden A M A, Van Veen J A and Frissel M J 1987 *Plant and Soil* **101** 21–8
[17] Tate R L 1991 *Soil Organic Matter: Biological and Ecological Effects* (New York: John Willey & Sons)
[18] Chertov O G, Bykhovets S S, Nadporozhskaya M A, Komarov A S and Larionova A A 2007 Evaluation of the rates of transformation of soil organic matter in the ROMUL model *Modellig of soil organic matter dynamics in forest ecosystems* ed V N Kudeyarov (Moscow: Nauka) 83–9
[19] Bohn T J, Podest E, Schroeder R, Pinto N, McDonald K C, Glagolev M, Filippov I, Maksyutov S, Heimann M, Chen X and Lettenmaier D P 2013 *Biogeosciences* **10-10** 6559–76

IOP Conf. Series: Earth and Environmental Science **138** (2018) 012005 doi:10.1088/1755-1315/138/1/012005

# Methane and carbon dioxide fluxes in the waterlogged forests of south and middle taiga of Western Siberia

**M V Glagolev*[1,2,3,4], D V Ilyasov[1], I E Terentieva[2], A F Sabrekov[2,4], S Yu Mochenov[3] and S S Maksutov[2,5]**

[1]Institute of Forest Science of the Russian Academy of Sciences, 143030, Moscow Region., Uspenskoe, Sovetskaya street, 21, Russia
[2]Tomsk State University, 634050, Lenina prospect, 34a, Russia
[3]Lomonosov Moscow State University, Faculty of Soil Science, 119991, Moscow, Leninskiye Gory, 1, Bldg. 12, Russia
[4] Yugorsky State University, 628012 Khanty-Mansiysk, Chehova street, 16, Russia
[5]National Institute for Environmental Studies, Onogawa, 16-2, Tsukuba, Ibaraki, 305-8506, Japan

*Email: m_glagolev@mail.ru

**Abstract.** Field measurements of methane and carbon dioxide flux were carried out using portable static chambers in south (ST) and middle taiga subzones (MT) of Western Siberia (WS) from 16 to 24 August 2015. Two sites were investigated: Bakchar bog in the Tomsk region (in typical ecosystems for this area: oligotrophic bog/forest border and waterlogged forest) and Shapsha in Khanty-Mansiysk region (in waterlogged forest). The highest values of methane fluxes ($mgC \cdot m^{-2} \cdot h^{-1}$) were obtained in burnt wet birch forest (median 6.96; first quartile 3.12; third quartile 8.95). The lowest values of methane fluxes (among the sites mentioned above) were obtained in seasonally waterlogged forests (median -0.08; first and third quartiles are -0.14 and -0.03 $mgC \cdot m^{-2} \cdot h^{-1}$ respectively). These data will help to estimate the regional methane flux from the waterlogged and periodically flooded forests and to improve its prediction.

## 1. Introduction.

Soils play an important role in the balance of the most important greenhouse gases – carbon dioxide ($CO_2$) and methane ($CH_4$). On the one hand, the soils accumulate carbon through plants photosynthesis. On the other hand, several soil ecosystems (mainly wetlands) are one of the most important natural $CH_4$ sources. The ecosystem classification may help to predict total greenhouse gas budget, since the quantitative contribution of different ecosystems in $CO_2$ and $CH_4$ exchange is still undefined and is discussed by the experts [1, 2].

Significant efforts were made to quantify the different soil $CH_4$ sources such as mires, lakes, rice fields and landfills [3, 4, 5, 6]. However, from a theoretical point of view, the $CH_4$ emission may occur from any ecosystems with excessive water supply (for example, waterlogged forests and floodplains). Unfortunately, the data on the methane emission from these ecosystems are insufficient. Mostly such studies were carried out in tropics [7, 8, 9], but for other regions there is a lack of these data [10, 11, 12]. Some useful information is given in reviews [3, 13]. Nevertheless, there are not enough data for global, or at least, regional assessments. Some forests are flooded only on a certain interval of the season, being a source of $CH_4$ during these intervals, but the rest part of the season they will consume it from the atmosphere. It makes regional methane flux assessment more difficult. The same is relevant to the river floodplains.

The aim of this paper is to present preliminary results of $CH_4$ and $CO_2$ flux field measurements in WS south and middle taiga waterlogged forests (WF) obtained at two sites: close to the field station "Plotnikovo" (Bakchar district of Tomsk oblast) of the Institute of Soil Science and Agrochemistry of SB RAS and to the field station "Shapsha" (Khanty-Mansiysk Autonomous Okrug) of Yugorsky State University.

## 2. Materials and methods.

The $CO_2$ and $CH_4$ flux measurements were carried out in August 2015 in WS south and middle taiga subzones. In the south taiga subzone (near Plotnikovo, Tomsk oblast; indicated in Figure 1 as "Transect") four measurement sites were located on a transect from the open oligotrophic bog with dominance of pine (*Pinus sylvestris)* and mosses (*Sphagnum sp.*) to the border of waterlogged forest with dominance of birch (*Betula pendula*).

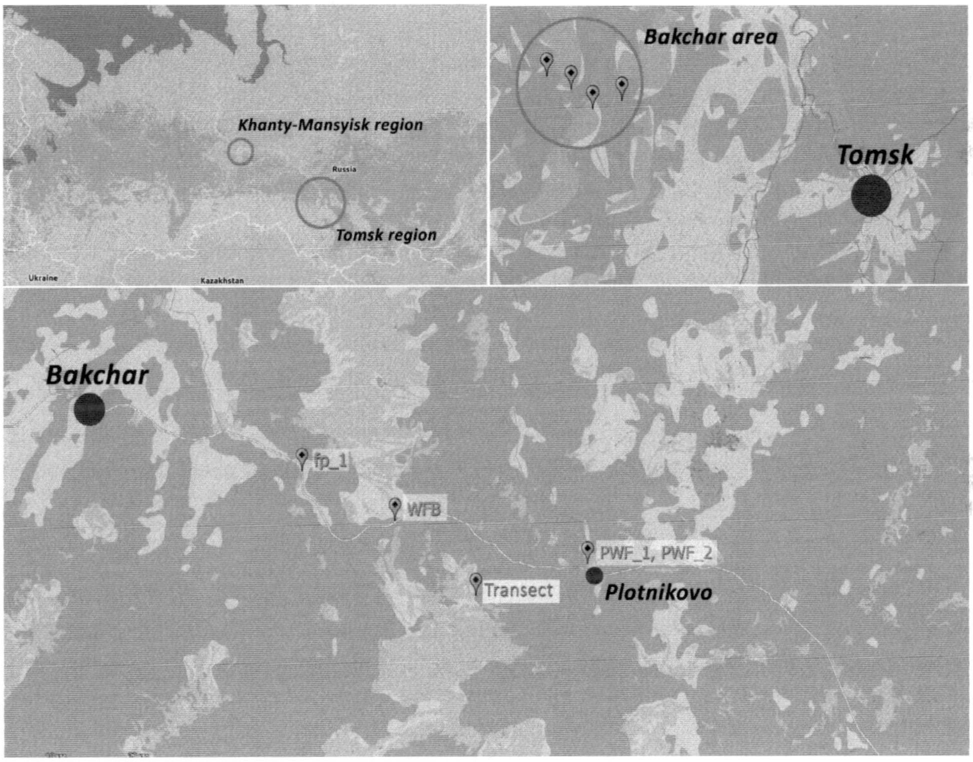


**Figure 1.** Location of test sites.

Thus, a wide range of soil moisture conditions and plant associations was studied. If possible the chambers were installed at hillocks and hollows (two points) at each microsite during the measurements. The first measurement point (Tr.PWF, 56°49.86800' N 82°51.16700' E) was located in a wet monodominant birch forest. The second one (Tr.WF/RB_2.1 and WF/RB_2.2, 56°49.88917' N 82°51.08000' E) and the third one (Tr.WF/RB_1, 56°49.90167' N 82°51.07333' E) in birch forest near the oligotrophic bog ("Bakchar wetland") border with admixture of pine (*Pinus sylvestris*). The fourth point (Tr.Ryam, 56°49.91667' N 82°51.04500' E) was located in oligotrophic bog with admixture of birch (*Betula pendula*). The water table level (WTL) ranged from 21 to 46 cm below the surface of soil, the soil pH ranged between (4.2-5.2), the water conductivity did not exceed 100 μS/cm. In addition, the measurements were carried out in periodically wet birch-spruce forest (PWF_1.1, 56°51.74400' N, 83°4.28200' E and PWF_1.2, 56°51.74400' N, 83°4.27900' E) and in the burnt wet birch forest (WFB site, 56°54.596' N, 82°41.811' E), where the WTL was (-20 cm) above the soil surface. Note that the Bakchar wetland ecosystem has been studied in the last years and described in detail (see e.g., [14, 15, 16, 17]).

The studied middle taiga forest located in the central part of the first terrace above the floodplain of the river's Ob and formed mainly by spruce (*Picea obovata*) with minor admixture of birch and aspen (trees height near 20-25 m). Excessive water supply is caused by poor drainage and high precipitation/evapotranspiration ratio. There is a pronounced microrelief such as flooded depressions and dry elevations formed by tussocks of *Sphagnum magellanicum*. The depth of organic horizon does not exceed 0.5 m; the water pH is about 3.9; the electrical conductivity is 200 μS/cm.

$CH_4$ and $CO_2$ fluxes were measured by static chamber method [18, 19]. The chamber was small (size of 0.4 × 0.4 × 0.4 $m^3$) portable opaque plexiglas cube without the lower side with forced ventilation inside. The chamber was installed on a steel square collar (0.4×0.4 $m^2$) embedded into undisturbed soil before measuring. A water channel on the chamber rim acted as a lock against leaks into or out of the chamber. Gas sampling from the chamber was carried out using 12 ml (for $CH_4$) or 20 ml (for $CO_2$) syringes four times ($t_0$, $t_1$, $t_2$ and $t_3$) through a silicone tube entering the chamber through the rubber stopper hermetically installed in the hole of a chamber sidewall. The total exposure time ($t_3$) did not exceed 60 minutes (in case of $CH_4$ sampling) or 15 minutes (in case of $CO_2$); the vegetation inside the chamber remained intact. The analyses of $CO_2$ concentrations were made with infrared gas analyzer DX-6100 ("RMT Ltd", Russia) not later than six hours after air sampling from the chamber. The concentration of $CH_4$ was measured using a modified gas chromatograph "KhPM-4" (Chromatograph, USSR) equipped with a flame ionization detector (column length 1 m; diameter 5 mm; sorbent – Sovpol, 80-100 mesh; temperature of column 40 °C; hydrogen as a carrier gas with flow rate 5 ml/min). The chromatograph "Crystal-5000" (Chromatec, Russia) also was used (two flame ionization detectors; column length – 2 m; diameter – 2 mm; sorbent – Porapak Q, 80-100 mesh; column temperature 150 °C; nitrogen $N_2$ as the carrier-gas with flow rate 10 ml/min). The chromatograph was calibrated with "standard" gas mixtures (1.99±0.01, 5.00±0.01 and 9.84±0.01 ppmv methane in a synthetic air; National Institute for Environmental Studies, Japan). Following physical and chemical environmental parameters were measured simultaneously with the air sampling at the study sites: soil temperature at depths of 0, 5, 15, 45 cm (sensors "THERMOCHRON" iButton DS 1921G "DALLAS Semiconductor", US); electrical conductivity (ES) and soil water pH (Kelilong PHT-028 "Kelilong Electron", China) and soil moisture (by the gravimetric method [20]). The water electrical conductivity and pH were measured only in cases when the ground water table level (WTL) was not deeper than 50 cm below the surface (length of sampling tube). Botanical descriptions were made at the measurement sites. The fluxes were calculated by linear regression for $CH_4$ and $CO_2$ emission and by non-linear regression for $CH_4$ uptake [21]. Positive fluxes related to the emission from soil to the atmosphere and negative fluxes related to the emission from atmosphere to the soil. A positive value of WTL means the water level is below the soil surface, a negative value – above it.

## 3. Results and Discussion.

The median of all obtained $CH_4$ fluxes was 0.02 $mgC{\cdot}m^{-2}{\cdot}h^{-1}$ and the first and third quartiles were -0.03 and 0.36 $mgC{\cdot}m^{-2}{\cdot}h^{-1}$, respectively. The probability density distribution of $CH_4$ fluxes is shown in Figure 2. The WTL values varied from -20 to 46 cm (median is 0 cm). $CH_4$ fluxes on "transect", as expected, changed from consumption (Tr.PWF) to small emission (Tr.Ryam) (Table 1). The median of $CH_4$ flux at a plot Tr.PWF was (-0.08 ± 0.06) $mgC{\cdot}m^{-2}{\cdot}h^{-1}$, it is comparable with PWF_1(2) (-0.02 ± 0.05) $mgC{\cdot}m^{-2}{\cdot}h^{-1}$ (Table 2, periodically waterlogged spruce forest). The median of $CH_4$ flux on a plot Tr.Ryam was (0.30 ± 0.05) $mgC{\cdot}m^{-2}{\cdot}h^{-1}$ with changing from 0.03 in elevations (WTL = 56 cm) to 0.50 $mgC{\cdot}m^{-2}{\cdot}h^{-1}$ in depressions (WTL=37 cm). The highest value of the $CH_4$ flux (6.96±0.74) $mgC{\cdot}m^{-2}{\cdot}h^{-1}$ was observed in the burnt birch forest (WFB) (Table 2; located outside the site "transect") and probably associated with high groundwater level (WTL = -20 cm).

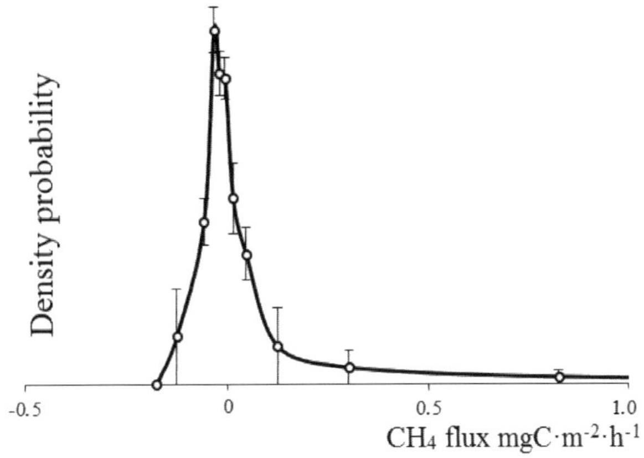


**Figure 2.** Probability density distribution of $CH_4$ fluxes (all obtained fluxes).

In the middle taiga spruce forest, when soil moisture was high, $CH_4$ fluxes were changed from (0.08 ± 0.04) to (1.20 ± 0.05) $mgC\cdot m^{-2}\cdot h^{-1}$ when WTL was changed from 0 to -5 cm, median is 0.46 $mgC\cdot m^{-2}\cdot h^{-1}$ and 1st and 3rd quartiles are 0.25 and 0.82 $mgC\cdot m^{-2}\cdot h^{-1}$, respectively (Table 3).

It is known that the $CH_4$ fluxes strongly correlate with the soil moisture [11]. Therefore, the obtained small values of $CH_4$ fluxes are expectable. Firstly, the WTL was significantly below the soil surface for the majority of the investigated ecosystems, resulting in the inhibition of $CH_4$ formation and favorable conditions for its consumption. Secondly, the soil moisture in the periodically flooded forests is less stable than in the wetland due to the smaller height of peat horizon, which stores water. The recovery of the methanogenic activity is slow after the abrupt change of hydrological conditions (during subsequent periods of flooding and drought).

According to the literature data, $CH_4$ fluxes in wet forests (Table 4) may range from -0.67 to 17.1 $mgC\cdot m^{-2}\cdot h^{-1}$ depending on soil moisture and other factors (microtopography, season, type of ecosystems). For comparison, the $CH_4$ flux in the dry upland forests ranges from -2.26 ± 0.17 to -0.02 $mgC\cdot m^{-2}\cdot h^{-1}$ [11, 23, 24, 25, 26].

Moss-grass and soil respiration (MgSR), i.e. soil $CO_2$ fluxes without photosynthesis, ranged from (174±32) to (414±142) $mgC\cdot m^{-2}\cdot h^{-1}$ on a site (Tr.WF/RB_1). It should be noted, that the correlation between the $CH_4$ fluxes and MgSR was negative (Figure 3). The highest mean $CH_4$ flux was (6.96±0.7) $mgC\cdot m^{-2}\cdot h^{-1}$ on the site WFB where the mean MgSR was the lowest (68.6±8.9 $mgC\cdot m^{-2}\cdot h^{-1}$) (Table 3). On the contrary, on the site Tr.PWF the mean $CH_4$ flux was minimal (-0.08 ± 0.06 $mgC\cdot m^{-2}\cdot h^{-1}$), while the mean MgSR was the highest (414±142) $mgC\cdot m^{-2}\cdot h^{-1}$ (Table 3).

These values are in a good agreement with the published data. For example, in the spruce forest [28], in the similar ecosystems of the south taiga in the European part of Russia, MgSR measured in 1993 and 1995-1997 were respectively 207, 130, 217 and 104 $mgC\cdot m^{-2}\cdot h^{-1}$. For comparison, MgSR at the waterlogged spruce forest (Tr.PWF) with periodically excessive water supply had the same magnitude scope and ranged from (206 ± 6) to (533 ± 21) $mgC\cdot m^{-2}\cdot h^{-1}$. This difference can be explained by more intensive respiration of the soil and vegetation in the south taiga, by the small number of our measurements and inter-annual variability of the $CO_2$ emission.

The MgSR values from our sites do not differ considerably from the values obtained in tropical forests. For example, [8] presents data for forests in the basins of the river Congo and Ubangi (Central Africa): (110 ± 57) $mgC\cdot m^{-2}\cdot h^{-1}$ – for flooded soils, (93 ± 11) $mgC\cdot m^{-2}\cdot h^{-1}$ – for wet forests and (80 ± 9) $mgC\cdot m^{-2}\cdot h^{-1}$ – for dry forests. There are works of other researchers (cited in [8]) where MgSR is 108 and 176 $mgC\cdot m^{-2}\cdot h^{-1}$ for rain forests of the Amazon region.

**Table 1.** Methane and carbon dioxide fluxes in "transect" site (Bakchar wetland, south taiga subzone).

| Temperature (°C)[a] | | | | | Flux ± error[b], mgC·m$^{-2}$·h$^{-1}$ | | WTL[e], cm | pH | EC[f], µS/cm |
|---|---|---|---|---|---|---|---|---|---|
| air | Depth from soil surface, cm | | | | | | | | |
| | 0 | 5 | 10 | 15 | $CH_4$ | $CO_2$ | | | |
| plot Tr.PWF, birch forest with excessive water supply, (date: 9.08.2015), dominant: *Betula pendula* | | | | | | | | | |
| 15.4/14.8 | 15.3/15.0 | 13.5/13.6 | 13 | 12.8/13.0 | -0.041±0.045(a) | 379±35(c) | | | |
| 15.5/14.9 | 15.9/15.3 | 14.5 | 13 | 13.3/13.4 | -0.111±0.057(a) | 290±22(c) | *172(0)* | Water table | |
| 14.8/15.1 | 14.6/14.7 | 13.5/13.5 | 13 | 12.5 | -0.150±0.135(a) | 489±41(c) | *353(5)* | was too deep | |
| 15.0/15.3 | 15.0/15.1 | 14.5 | 13 | 13.5/13.3 | -0.031±0.028(a) | 273±14(c) | *454(10)* | to take samples | |
| 14.4/14.5 | 14.3/14.4 | 13.6/13.5 | 13.2/13.0 | 12.9/12.6 | -0.164±0.048(a) | 509±26(c) | *503(15)* | for analysis | |
| 14.5 | 14.4/14.5 | 14.3/14.5 | 13.2/13.0 | 13.5/13.5 | -0.031±0.070(a) | 226±9(c) | | | |
| – "– (date: 16.08.2015) | | | | | | | | | |
| 18.7/18.9 | 16.2/16.5 | 15 | 12.5 | 13 | -0.073±0.190(a) | 280±19(c) | n.d. | | |
| 18.7/18.9 | 16.2/16.5 | 15 | 12.5 | 13 | -0.092±0.040(a) | 399±122(c) | n.d. | Water table | |
| 18.7/19.0 | 17 | 15.5/15.5 | 13 | 12.5 | -0.040±0.024(a) | 175±5(b, c) | n.d. | was too deep | |
| 20.2/19.6 | 17.2/16.7 | 15.9/15.1 | 13.2/12.5 | 13.6/13.0 | -0.028±0.084(a) | 566±181(c) | n.d. | to take samples | |
| 20.2/19.7 | 17.2/16.8 | 15.9/15.2 | 13.2/12.5 | 13.6/13.0 | -0.153±0.250(a) | 471±208(c) | n.d. | for analysis | |
| 20.4/19.9 | 17.9/17.5 | 15.9/15.5 | 13.8/13.0 | 13.2/12.5 | -0.085±0.006(a) | 429±163(c) | n.d. | | |
| plot Tr.WF/RB_1, the periodically wet forest at the border of the pine-shrub-sphagnum community,(date: 16.08.2015) | | | | | | | | | |
| 20.4/19.5 | 18.2/18.0 | 15.6/15.8 | 13.5/13.6 | 13 | 0.260±0.051(b) | 374±20(c) | 26.5 | 4.7 | 70 |
| 20.4/19.5 | 18.2/18.0 | 15.6/15.8 | 13.5/13.6 | 13 | 1.322±0.579(c) | 491±16(c) | 26.5 | 4.7 | 70 |
| 20.4/19.6 | 19.1/19.0 | 14.0/14.1 | 13.0/13.1 | 12.5 | 0.164±0.058(c) | 172±48(c) | 26.5 | 4.7 | 70 |
| 19.6/19.9 | 17.8/18.0 | 15.7/15.5 | 13.9/13.5 | 13.4/13.0 | 0.025±0.047(b) | 304±28(c) | 26.5 | 4.7 | 70 |
| 19.6/19.9 | 17.7/18.0 | 15.8/15.5 | 14.0/13.5 | 13.5/13.0 | 0.775±0.334(c) | 479±97(c) | 26.5 | 4.7 | 70 |
| 19.6/19.9 | 18.1/18.5 | 14.8/14.2 | 13.6/13 | 13.1/12.5 | 0.006±0.060(c) | 338±11(c) | 26.5 | 4.7 | 70 |
| – "– (date: 18.08.2015) | | | | | | | | | |
| 25.6/25.8 | 20.6/20.8 | 22.0/21.7 | 21.2$^{c}$/21.0$^{c}$ | 10.7$^{c}$/11.0$^{c}$ | -0.017±0.107(a) | 331±79(c) | 21 | 4.4 | 40 |
| 25.5/25.8 | 20.6/20.8 | 22.0/21.7 | 21.2$^{c}$/21.0$^{c}$ | 10.7$^{c}$/11.0$^{c}$ | -0.022±0.018(a) | 149±32(c) | 21 | 4.4 | 40 |
| 25.5/25.8 | 21.5/21.8 | 17.5/18.0 | 13.1$^{c}$/13.2$^{c}$ | 10.6$^{c}$/10.9$^{c}$ | 0.050±0.049(b) | 622±152(c) | 21 | 4.4 | 40 |
| 24.8/24.6 | 20 | 23.0/22.9 | 22.1$^{c}$/22.0$^{c}$ | 10.53 | -0.029±0.081(a) | 174±31(c) | 21 | 4.4 | 40 |
| 24.8/24.5 | 20 | 23 | 22.1$^{c}$/22.0$^{c}$ | 10.53 | -0.042±0.080(a) | 130±24(c) | 21 | 4.4 | 40 |
| 24.8/24.5 | 21 | 17 | 13.03 | 10.53 | -0.001±0.092(a) | 461±191(c) | 21 | 4.4 | 40 |
| plotsTr.WF/RB_2.1(2.2),the periodically wet forest near the border of the pine-shrub-sphagnum community,(date:08.08.2015) | | | | | | | | | |
| 23.2/23.5 | 20.4/20.5 | 24 | 23.2$^{c}$/23.1 | 11.8$^{c}$/12.0$^{c}$ | 0.167±0.039(c) | 352±8(c) | 37 | 4.9 | 45 |
| 23.2/23.5 | 21.0/21.1 | 19.3/19.5 | 15.0$^{c}$/15.0 | 12.5$^{c}$/12.5$^{c}$ | 0.008±0.099(b) | 311±4(c) | 46 | 5.5 | 45 |
| 23.2/23.5 | 21.0/21.1 | 19.3/19.5 | 15.0$^{c}$/15.0 | 12.5$^{c}$/12.5$^{c}$ | 0.060±0.032 | 236±81(c) | 46 | 5.5 | 45 |
| 22.0/22.5 | 19.7/20.0 | 23.9/24.0 | 23.0$^{c}$/23.0 | 11.53 | 0.079±0.046(b) | 349±43(c) | 37 | 4.9 | 45 |
| 21.9/22.4 | 20.18/20.5 | 19 | 15.0$^{c}$/15.0 | 12.1$^{c}$/12.3$^{c}$ | -0.017±0.089(a) | 50±48(c) | 46 | 5.5 | 45 |
| 20.8/21.0 | 19 | 23.4/23.5 | 23.0$^{c}$/23.0 | 11.53 | 0.023±0.065(b) | 213±46(c) | 37 | 4.9 | 45 |
| 20.8/21.0 | 19.43/19.5 | 18.5 | 15.0$^{c}$/15.0 | 12.03 | -0.065±0.029(a) | 441±90(c) | 46 | 5.5 | 45 |
| 20.8/21.0 | 19.43/19.5 | 18.5 | 15.0$^{c}$/15.0 | 12.03 | 0.134±0.106(b) | 322±8(c) | 46 | 5.5 | 45 |
| – "– (date: 22.08.2015) | | | | | | | | | |
| 12.0/11.2 | 12.0/11.2 | 13.0/13.1 | 14.0$^{c}$/14.0 | 11.8$^{c}$/12.0$^{c}$ | 0.052±0.053(b) | 630±299(c) | n.d. | n.d. | n.d. |
| 12.1/11.2 | 12.1/11.2 | 13.2/13.7 | 14.3$^{c}$/14.5 | 12.5$^{c}$/12.5$^{c}$ | -0.021±0.018(a) | 367±135(c) | n.d. | n.d. | n.d. |
| 12.1/ n.d. | 12.1/ n.d. | 13.2/ n.d. | 14.3$^{c}$/ n.d. | 12.5$^{c}$/ n.d. | 0.036±0.016(b) | n.d. | n.d. | n.d. | n.d. |
| 12.5/12.3 | 12.42/12.1 | 12.9/13,0 | 13.4$^{c}$/13.5 | 11.53 | 0.033±0.021(c) | 299±36(c) | n.d. | n.d. | n.d. |
| 12.5/12.3 | 11.5/11.5 | 12.9/13,0 | 14.0$^{c}$/14.0 | 12.53 | -0.029±0.054(a) | 200±89(c) | n.d. | n.d. | n.d. |
| 12.5/12.3 | 11.5/11.5 | 12.9/13,0 | 14.0$^{c}$/14.0 | 12.53 | -0.003±0.072(a) | 190±25(c) | n.d. | n.d. | n.d. |
| plot Tr.Ryam, pine-shrub-sphagnum community, (date: 22.08.2015), dominant: *Pinus sylvestris* | | | | | | | | | |
| 14.6/13.6 | 14.1/14.4 | 14 | 12.03 | 9.53 | 0.359±0.017(c) | 194±96(c) | n.d. | n.d. | n.d. |
| 14.6/13.6 | 14.1/14.4 | 14 | 12.03 | 9.53 | 0.495±0.184(c) | 180±12(c) | n.d. | n.d. | n.d. |
| 14.8/13.5 | 13.6/13.8 | n.d. | 13.53 | 10.7$^{c}$/11.0$^{c}$ | 0.162±0.036(b) | 181±8(b) | n.d. | n.d. | n.d. |
| 13.7/ n.d. | 13.0/ n.d. | 13.5/ n.d. | 12.0$^{c}$/ n.d. | 9.0$^{c}$/ n.d. | 0.245±0.055(b) | n.d. | n.d. | n.d. | n.d. |
| 13.7/14.2 | 13.0/13.1 | 13.5 | 12.0$^{c}$/12.0$^{c}$ | 9.03 | 0.503±0.208(c) | 78±12(b) | n.d. | n.d. | n.d. |
| 13.7/14.3 | 13.0/13.2 | 13.9/14.0 | 13.4$^{c}$/13.5$^{c}$ | 10.53 | 0.031±0.050(c) | 186±46(c) | n.d. | n.d. | n.d. |

[a, b, c, d, e] see notes of table 2.

IOP Conf. Series: Earth and Environmental Science **138** (2018) 012005 doi:10.1088/1755-1315/138/1/012005

**Table 2.** Methane and carbon dioxide fluxes in the forests with different soil moisture (south taiga).

| | Temperature (°C)[a] | | | | Flux ± error[b], $mgC \cdot m^{-2} \cdot h^{-1}$ | | WTL[e], cm |
|---|---|---|---|---|---|---|---|
| air | soil depth | | | | | | |
| | 0 cm | 5 cm | 10 cm | 15 cm | $CH_4$ | $CO_2$ | |
| PWF_1.1 and PWF_1.2, birch-spruce periodically waterlogged forest, site "Plotnikovo" (date: 5.08.2015), dominant species: *Betula pendula, Sorbus sibirica* | | | | | | | |
| 15.9 | 17 | 16.1 | 16.7 | n.d. | -0.013±0.193(a) | 206±6(b) | n.d. |
| 15.9 | 17 | 16.1 | 16.7 | n.d. | -0.023±0.060(a) | 227±15(c) | n.d. |
| 15.9 | 17 | 16.1 | 16.7 | n.d. | -0.065±0.081(a) | 403±121(c) | n.d. |
| 15.9 | 17 | 16.1 | 16.7 | n.d. | -0.012±0.057(a) | 428±36(c) | n.d. |
| 18.4 | 18.6 | 16.5 | 16.5 | n.d. | 0.016±0.037(c) | 227±33(c) | n.d. |
| 18.4 | 18.6 | 16.5 | 16.5 | n.d. | -0,064±0,031(a) | 509±280(c) | n.d. |
| 18.4 | 18.6 | 16.5 | 16.5 | n.d. | -0.001±0.070(a) | 351±15(c) | n.d. |
| 18.4 | 18.6 | 16.5 | 16.5 | n.d. | 0.053±0.035(b) | 315±10(c) | n.d. |
| 15.5 | 16.7 | 16.5 | 16.5 | n.d. | 0.061±0.043(c) | 353±11(c) | n.d. |
| 15.5 | 16.7 | 16.5 | 16.5 | n.d. | -0.041±0.063(a) | 533±21(c) | n.d. |
| 15.5 | 16.7 | 16.5 | 16.5 | n.d. | -0.028±0.048(a) | 471±31(c) | n.d. |
| 15.5 | 16.7 | 16.5 | 16.5 | n.d. | -0.065±0.058(a) | 467±29(c) | n.d. |
| WFB, wet burnt birch forest, site "Bakchar bog" (date: 24.08.2015), dominant species: *Betula péndula, Calla palustris* | | | | | | | |
| 17.9/18.8 | 14.6/14.9 | 14.5 | $12.5^c/12.6^c$ | $11.6^c/11.8^c$ | 6.839±1.155(b) | 42±19(c) | -20 |
| 17.9/18.8 | 14.6/14.9 | 14.5 | $12.5^c/12.6^c$ | $11.6^c/11.8^c$ | 3.341±0.404(c) | 40±10(c) | -20 |
| 17.9/18,8 | 15.3/15.5 | 14.8/15.0 | 13.6/13.8 | 11.5 | 9.419±3.442(c) | 174±23(c) | -20 |
| 17.3/17.1 | 13.8/14.0 | 14 | 12.1/12.2 | 11.5 | 8.395±1.642(a) | 152±41(a) | -20 |
| 17.3/17.1 | 13.8/14.0 | 14 | 12.1/12.2 | 11.5 | 1.599±0.132(c) | 132±56(c) | -20 |
| 17.3/17.1 | 14.7/14.9 | 14.5/14.5 | 13.5/13.5 | 11.5 | 10.936±0.762(b) | 116±6(c) | -20 |
| 17.3/17.1 | 14.7/14.9 | 14.5/14.5 | 13.5/13.5 | 11.5 | 8.945±1.499(c) | 127±89(b) | -20 |
| 14.5/15.1 | 13.2 | 13.8/14 | 12 | 11.5 | 2.204±0.152(c) | 25±8(c) | -20 |
| 14.5/ n.d. | 13.2/ n.d. | 13.8/ n.d. | 12.0/ n.d. | 11.5/ n.d. | 5.623±0.356(c) | n.d. | -20 |
| 14.5/15.1 | 13.8/14.1 | 14.4/14.5 | 13.5/13.5 | 11.5 | 6.987±0.71 | 57±7(a) | -20 |
| 14.5/15.1 | 13.8/14.1 | 14.4/14.5 | 13.5/13.5 | 11.5 | 9.699±2.36(c) | 118±48(c) | -20 |
| 10.1/11.6 | 12.5/12.9 | 13.2/13.5 | 12 | 11.5 | 7.174±0.851(c) | 50±7(c) | -20 |
| 10.1/11.6 | 12.5/12.9 | 13.2/13.5 | 12 | 11.5 | 2.473±0.151(c) | 80±41(c) | -20 |
| 10.1/11.6 | 13.0/13.3 | 13.9/14 | 13.5 | 11.5 | 6.925±2.157(b) | 87±5(c) | -20 |
| 10.1/ n.d. | 13.0/ n.d. | 13.9/ n.d. | 13.5/ n.d. | 11.5/ n.d. | 8.971±0.666(c) | 174±23(b) | -20 |
| fp_1, birch forest, site "Plotnikovo" (date: 14.08.2015), dominant species: *Betula pendula* | | | | | | | |
| 19.4/19.5 | 16.6/16.5 | 15.5/15.6 | 14.8/15.0 | 13.5 | 0.130±0.060(a) | 298±14(c) | *51(0)* |
| 19.4/19.5 | 16.6/16.5 | 15.5/15.6 | 14.8/15.0 | 13.5 | 0.130±0.060(a) | 298±14(c) | *38(5)* |
| 18.9/19.0 | 16.4/16.5 | 15.5/15.5 | 14.5 | 13.5 | 0.053±0.045(a) | 300±10(c) | *35(10)* |
| 18.2/18.5 | 16.0/16.2 | 15.4/15.5 | 14.5 | 13.5 | 0.070±0.087(a) | 271±96(c) | *35(15)* |

[a] the temperature during the $CH_4$ flux measurement / the temperature during the $CO_2$ flux measurement (under the same temperature there was only one value);

[b] types of error: (a) – confidence interval at 95%; (b) – combined error, calculated according to [22]; (c) – standard deviation; (d) - standard deviation calculated with the weights which are inversely proportional to a variance of gas concentration;

[c] soil temperature measured at the depth of 45 cm;

[d] soil temperature measured at the depth of 25 cm;

[e] groundwater level WTL (positive values – below ground level, negative – above); in case where the values in this column are given in italics – it is a soil moisture (%) at depth (cm) specified in parenthesis (not WTL!);

[f] EC – electrical conductivity.

**Table 3.** $CH_4$ fluxes at "Shapsha" site (Sh.WFor.1.1-1.10, middle taiga, 27.8.2015).

| Temperature (°C)[a] | | | | flux± error[b], $mgC\cdot m^{-2}\cdot h^{-1}$ | WTL[e], cm | pH |
|---|---|---|---|---|---|---|
| air | Soil depth | | | | | |
| | 5 cm | 10 cm | 25 cm | | | |
| Spruce waterlogged forest, dominant species: *Picea obovata*, *Carex* sp., *Sphagnum magellanicum* | | | | | | |
| 11.1 | 9.5 | 9.0 | 8.5 | 0.238±0.017(d) | -3 | 3.9 |
| 11.1 | 9.5 | 9.0 | 8.5 | 0.600±0.022(d) | -3 | 3.9 |
| 10.6 | 9.4 | 8.8 | 8.4 | 0.081±0.042(d) | -3 | 3.9 |
| 10.6 | 9.4 | 8.8 | 8.4 | 0.890±0.008(d) | -3 | 3.9 |
| 10.3 | 9.3 | 8.8 | 8.4 | 0.235±0.017(d) | -5 | 3.9 |
| 10.3 | 9.3 | 8.8 | 8.4 | 0.563±0.030(d) | -5 | 3.9 |
| 9.6 | 9.0 | 8.8 | 8.4 | 1.152±0.025(d) | 0 | 3.9 |
| 9.6 | 9.0 | 8.8 | 8.4 | 0.302±0.061(d) | 3 | 3.9 |
| 8.8 | 8.7 | 8.8 | 8.4 | 1.201±0.051(d) | 0 | 3.9 |
| 8.8 | 8.7 | 8.8 | 8.4 | 0.362±0.026(d) | 3 | 3.9 |

[a, b, e] see notes to table 2.

**Table 4.** $CH_4$ flux in forests with excessive water supply.

| Ecosystem | $CH_4$ flux, $mgC\cdot m^{-2}\cdot h^{-1}$ | Source | Note |
|---|---|---|---|
| Flooded forests in the Amazon river basin. | 3.4<br>6.0 | Devol et al.<br>Bartlett et al. | Data published in 1988 [8] |
| Forests in Central Africa in the Congo and Oubangui river basin | 0.3 ÷ 17.1<br>0.04 ÷ 0.24<br>-0.03 ÷ -0.14 | [8] | Flooded forests (WTL from -10 to -40 cm)<br>Forests on moist soils (WTL from 10 to 20 cm)<br>Forests on drained soils (WTL higher 1 m) |
| Forests in Central Africa to the SW and W from the Impfondo | 0.19 or 0.67 (depends on method) | [8] | Measurements were performed in two variants of the gradient method (footprint ~ a few hundreds of $m^2$); fraction of flooded soil~1/3 |
| Forests in Central Africa to the NE and N from Brazzaville | 3.2 ÷ 6.5<br>2.4 ÷ 4.9 | [8] | During the wet season (average from territory >> $n\cdot 10^2$ $m^2$)<br>During the dry season (average from territory >> $n\cdot 10^2$ $m^2$) |
| Forest in Puerto Rico (18°18' N, 65°50' W) | 3.1 ± 1.6[a]<br>0.010 ± 0.008[a]<br>-0.015 ± 0.002[a] | [8] | *Tabebuia rigida* forest<br>*Cyrilla racemiflora* forest<br>*Dacryodes excelsa* forest |
| Spruce (*Picea abies L.*) forest, Denmark | from -0.001±0.005[a] to -0.030±0.004[a] | [10] | The data from Fig. 6 in the original publication |
| Pine forest (39°55'N, 74°35'W) | 0.032 ± 0.008[b]<br>-0.046 ± 0.007[b] | [11] | Elevation (WTL 7 m)<br>Depression (WTL ± 5 cm) |
| Floodplain Alder forest (periodically flooded) | from -0.028 to 0.025 | [27] | sites are located in Alaska, USA (64°45′N, 148°18′ W) |
| Floodplain Spruce Forest (periodically flooded). | -0.0121 ± 0.0008[c] | | |

[a] standard errors after «±».
[b] standard deviations after «±».
[c] there are no information about the type of error after «±».

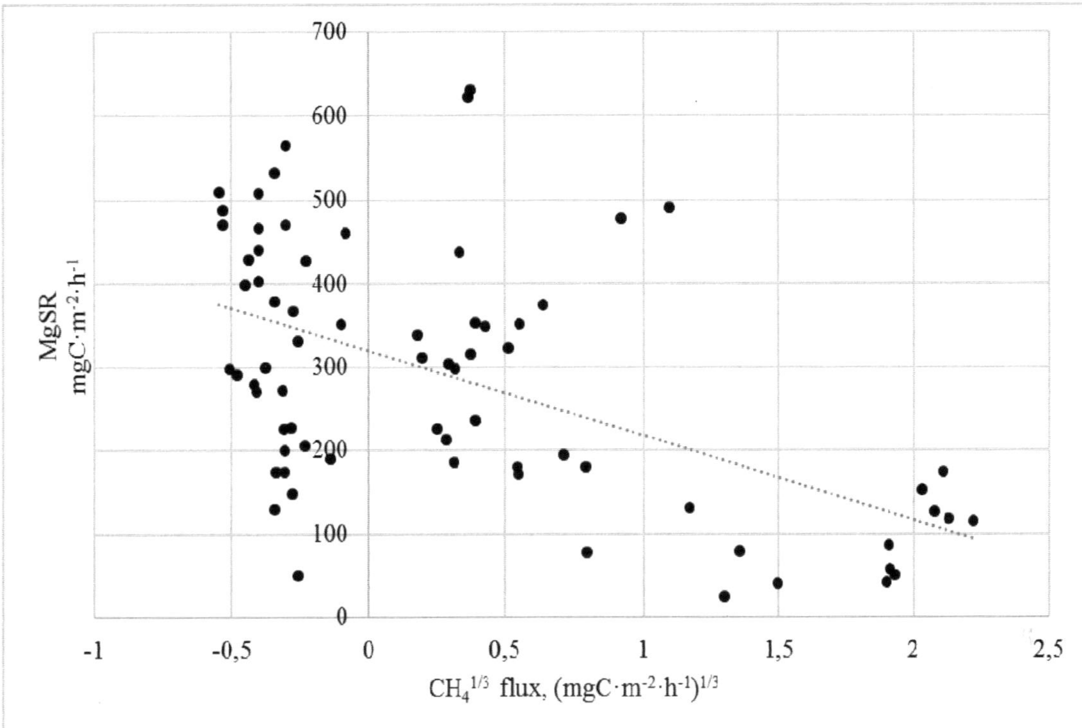


**Figure 3.** The relationship between $CH_4$ and $CO_2$ fluxes (without points Sh.WFor, where carbon dioxide fluxes were not measured). For better readability *x*-axis is not a $CH_4$ flux, but it is a cube root of it.

Ambus and Christensen [10] studied several ecosystems where temporary waterlogging was possible. They suggested the following important assumption: calculation of the total flux for the periodically waterlogging ecosystems should be performed with respect to the topography of the landscape. In this case, for correct estimation of gas flux using the chamber method it is necessary to take into account relative water levels during flooding (in addition to flux measurements at the flooded areas). Unfortunately, although these authors have studied waterlogged forests, in the end they did not include waterlogged forests in the list of ecosystems for which their assumption was relevant. Our results demonstrate both consumption and emission of methane in waterlogged forests. It allows us to extend this approach to forests, at least to those located on the border of wetlands.

To increase the accuracy of the flux prediction for these soils, it is necessary to make measurements with the highest possible spatial and temporal resolution [10]. Indeed, during a single measurements session in a season, the emission may be zero, but it does not mean that this site does not emit methane during the season. Apparently, if providing a detailed (in space and time) flux data is not possible, calculations may be a useful option. Knowing the topography and hydrology of for each point of a certain area it is possible to reveal how long and how often this point is relatively wet or dry.

**Acknowledgments**
Authors are deeply thankful to the head of "Plotnikovo" field station (ISSA SB RAS) Dr. Smolentsev B.A. and director of the ISSA SB RAS Prof. Syso A.I. for kindly providing the opportunity to carry out our work.

This work was supported by the RFBR grants 15-05-07622 and 15-44-00091, and BIO-GEO-CLIM № 14.B25.31.0001. S. Maksyutov was supported by the grant 2A1202 Environment Research and Technology Development Fund (Ministry of the Environment, Japan).

IOP Conf. Series: Earth and Environmental Science **138** (2018) 012005 doi:10.1088/1755-1315/138/1/012005

**References**

[1] Glagolev M V and Filippov I V 2015 A reply to A.V. Smagin: III. On the issue of methanotrophic filter and gas discharge into the atmosphere *Environmental Dynamics and Global Climate Change* **6** 42-54

[2] Kurganova I N and Kudeyarov V N 2015 Is it possible significant positive imbalance of carbon cycle (sink) in Russian territory? *Environmental Dynamics and Global Climate Change* **6** 32-35

[3] Bartlett K B and Harriss R C 1993 Review and assessment of methane emissions from wetlands *Chemosphere* **26** 261-320

[4] Boeckx P and van Cleemput O 1996 Flux estimates from soil methanogenesis and methanotrophy: landfills, rice paddies, natural wetlands and aerobic soils *Environ. Monit. Assess.* **42** 189-207

[5] Glagolev M V, Sabrekov A F, Kleptsova I E, Filippov I V, Lapshina E D, Machida T and Maksyutov Sh Sh 2012 Methane Emission from Bogs in the Subtaiga of Western Siberia: The Development of Standard Model *Eurasian Soil Sci.* **45** 947-57

[6] Sabrekov A F, Runkle B R K, Glagolev M V, Terentieva I E, Stepanenko V M, Kotsyurbenko O R, Maksyutov S S and Pokrovsky O S 2017 Variability in methane emissions from West Siberia's shallow boreal lakes on a regional scale and its environmental controls *Biogeosciences* **14** 3715-42

[7] Devol A H, Richey J E, Forsberg B R and Martinelli L A 1990 Seasonal Dynamics in Methane Emissions from the Amazon River Floodplain to the Troposphere *J. Geophys. Res.* **95** 16417-26

[8] Tathy J P, Cros B, Delmas R A, Marenco A, Servant J and Labat M 1992 Methane emission from flooded forest in Central Africa *J. Geophys. Res.* **97** 6159-68.

[9] Silver W L, Lugo A E and Keller M 1999 Soil oxygen availability and biogeochemistry along rainfall and topographic gradients in upland wet tropical forest soils *Biogeochemistry* **44** 301-28

[10] Ambus P and Christensen S 1995 Spatial and Seasonal Nitrous Oxide and Methane Fluxes in Danish Forest-, Grassland-, and Agroecosystems *J. of Environ. Qual.* **24** 993-1001.

[11] Aronson E L, Vann D R and Helliker B R 2012 Methane flux response to nitrogen amendment in an upland pine forest soil and riparian zone *J. Geophys. Res.* **117** G03012

[12] Berezina E V, Elansky N F, Moiseenko K B, Safronov A N, Skorokhod A I, Lavrova O V, Belikov I B and Shumsky R A 2014 Estimation of Biogenic $CH_4$ and $CO_2$ Emissions and Dry Deposition of $O_3$ using $^{222}$Rn Measurements in TROICA Expeditions *Izvestiya, Atmospheric and Oceanic Physics* **50** 583–94

[13] Glagolev M V 2010 Annotated reference list of $CH_4$ and $CO_2$ flux measurements from Russia mires *Environmental Dynamics and Global Climate Change* **1** 5-57

[14] Glagolev M V, Golovatskaya E A and Shnyrev N A 2008 Greenhouse Gas Emission in West Siberia *Contemp. Probl. Ecol.* **1** 136-46.

[15] Koronatova N G and Milyaeva E V 2011 The productivity of wetland pine forests in southern taiga of Western Siberia *Interexpo Geo-Siberia* **4** 269-73

[16] Sabrekov A F, Glagolev M V, Kleptsova I E, Machida T and Maksyutov S S 2013 Methane Emission from Mires of the West Siberian Taiga *Eurasian Soil Sci.* **46** 1182-93

[17] Mironycheva-Tokareva N P, Kosykh N P and Vishnykova E K 2013 Production and destruction processes in peatland ecosystems of Vasyugan region *Environmental Dynamics and Global Climate Change* **4** 1-9

[18] Nozhevnikova A, Glagolev M, Nekrasova V, Einola J, Sormunen K and Rintala J 2003 The analysis of methods for measurement of methane oxidation in landfills *Water Sci. Technol.* **48** 45-52

[19] Terent'eva I E, Sabrekov A F, Glagolev M V, Lapshina E D, Smolentsev B A and Maksyutov Sh Sh 2017 A new map of wetlands in the southern taiga of the West Siberia for assessing the emission of methane and carbon dioxide *Water Resources* **44** 297-307

[20] Dospekhov B A, Vasil'ev I P and Tulikov A M 1987 *Practical guide on agriculture* (Moscow: Agropromizdat)

[21] Sabrekov A F, Glagolev M V, Alekseychik P K, Smolentsev B A, Terentieva I E, Krivenok L A and Maksyutov S S 2016 A process-based model of methane consumption by upland soils *Environ. Res. Lett.* **11** 075001

[22] Glagolev M V and Smagin A V 2005 *Matlab Applications for Numerical Simulations in Biology, Ecology and Soil Science* (Moscow: Moscow St. Univ. Soil Sci. Dept.)

[23] Sullivan B W, Kolb T E, Hart S C, Kaye J P, Dore S and Montes-Helu M 2008 Thinning reduces soil carbon dioxide but not methane flux from southwestern USA ponderosa pine forests *Forest Ecol. Manage.* **255** 4047-55

[24] Castro M S, Steudler P A, Melillo J M, Aber J D and Bowden R D 1995 Factors controlling atmospheric methane consumption by temperate forest soils *Global Biogeochem. Cycles* **9** 1-10

[25] Groffman P M, Hardy J P, Driscoll C T and Fahey T J 2006 Snow depth, soil freezing, and fluxes of carbon dioxide, nitrous oxide and methane in a northern hardwood forest *Global Change Biol.* **12** 1748-60

[26] Hu R, Kusa K and Hatano R 2001 Soil respiration and methane flux in adjacent forest, grassland, and cornfield soils in Hokkaido, Japan *Soil Sci. Plant Nutr.* **47** 621-27

[27] Gulledge J and Schimel J P 2000 Controls on soil carbon dioxide and methane fluxes in a variety of taiga forest stands in interior Alaska *Ecosystems* **3** 269-82

[28] Glukhova T V, Kovalev A G, Smagina M V and Vompersky S E 1999 Assessment of some biotic components of the carbon cycle in forests and swamps *Wetlands and Waterlogged Forests in the Objectives of Environmental Sustainability. Materials of conf.* (Moscow: GEOS) 182-85

# Features of seasonal temperature variations in peat soils of oligotrophic bogs in south taiga of Western Siberia

**M V Kiselev**[1], **E A Dyukarev**[1], **N N Voropay**[1,2]

[1]Institute of Monitoring of Climatic and Ecological System of the Siberian Branch Russian Academy of Sciences, Akademicheskii 10/3, Tomsk, Russia
[2] V B Sochava Institute of Geography of the Siberian Branch Russian Academy of Sciences, St. Ulan-Bator, Irkutsk, Russia

Email: xplutox@yandex.ru, dekot@mail.ru, voropay_nn@mail.ru

**Abstract.** The work presents the results of the study of the peculiarities of the temperature regime in the five basic ecosystems of oligotrophic bogs in the south taiga zone of Western Siberia in 2011-2016. The soil temperature regime was studied using the atmospheric-soil measuring complex at different depths from surface down to 240 cm. All sites were divided into two groups according to the bog water level: flooded sites (hollow and open fen) and drained sites (ridge, tall and low ryam). The waterlogged sites are better warmed in the summer period and slowly freeze in the winter period. The analysis of the annual cycle of temperature showed that the maximum surface temperature is in July. The minimum temperature on the surface is observed in February or January. The greatest temperature gradient was recorded in the upper 2 cm layer. The gradient at the open fen was -2 °C/cm in February and 1.1 °C/cm in October. The peak of formation of the seasonally frozen layer occurs at the end of autumn or in the beginning of winter. The degradation of the seasonally frozen layer was observed both from top and bottom, but the degradation rate from the top is faster.

## 1. Introduction

The mires are unique natural landscapes involved in the regulation of the gaseous composition of the atmosphere, the water balance of the biosphere, biological diversity of the Earth [1, 2]. The soil temperature affects the growth of ground vegetation (mosses, sedges, shrubs, etc.). The soil microclimate is a key factor controlling many biotic and abiotic processes in the peat soils: decomposition and mineralization of soil organic matter [3], emissions of greenhouse gases ($CO_2$, $CH_4$, $N_2O$) [4], excretion of dissolved organic carbon [2].

The thermal regimes of peat and mineral soils are significantly different. Peat deposit is a complex organic system, which has specific properties: high water content and porosity composed of a large number of weakly decomposed organic matter [12]. The main storages of peat are located in Western Siberia. The peatland ecosystems of West Siberia contain up to 70 billion tons of carbon [11]. Such a large reservoir potentially may go into the atmosphere as CO2 or CH4 as the result of climate change or anthropogenic impacts and make a significant contribution to the carbon balance of the atmosphere.

The investigation of thermal regime of peat soils in the background of the global warming observed throughout the entire territory of Russia [9] is a vital scientific task. The purpose of this study is the analysis of soil temperature regime of typical bog ecosystems in the South Taiga of Western Siberia based on results of long-term monitoring.

## 2. Methods

The research of the temperature regime of soils was carried out in the geophysical scientific field station "Vasyuganje" at the Bakcharskoe bog (area of about 1400 $km^2$) [6, 10], located in the interfluves of the rivers Iksa and Bakchar in Bakcharsky district of Tomsk region, Russia [7]. Four typical bog ecosystems were chosen for soil temperature monitoring: pine-shrub-sphagnum ecosystem with trees height about 18 m (tall ryam), ryam with oppressed tree stand 2-3 m (low ryam), open sedge-sphagnum fen and ridge-hollow complex (RHC). The observation sites at RHC were organized both at oligotrophic ridge and sheizeria-sphagnum hollow [2].

The measurements of soil temperature were performed at several depths from surface to 240 cm (0, 2, 5, 10, 15, 20, 30, 40, 60, 80, 120, 160, 240 cm), and the air temperature was measured at a height of 2 m using the atmospheric-soil measuring complex [8]. The measurements were carried out from 1 April 2011 till 30 April 2016 (during 1858 days) with a time step of 15 minutes. The data on snow depth and air temperature from the nearest weather station "Bakchar" were derived from the archive (http://www.meteo.ru/).

The depth of peat freezeing was determined by the penetration depth of 0 °C temperature into the ground [5]. The depth of freezing was calculated by linear interpolation of the soil temperatures between two adjacent layers, providing that the temperature in one of the layers was negative.

## 3. Results and discussions

The analysis of annual temperature of peat soil in the profile 0-240 cm can identify a number of differences between the observation sites. Despite the fact that the sites located in the 300-1000 m from each other, and that the density and moisture content of the peat at the sites varies little, all sites may be divided in two groups. The sites of the first group warmed up better during the warm period than the sites of the second group. The first group includes sites with water level located at 3-7 cm from the surface (hollow at RHC and open fen). The water level at sites from the second group (tall and low ryam, ridge at RHC) decreases to 20-30 cm in the end of summer. While the maximum annual surface temperature in July on hollow and open fen was 18.1 and 17.4 °C respectively, for low ryam, tall ryam and ridge of RHC it was 17.3, 14.1 and 17.7 °C respectively.

The soil temperature in the warm period is associated not only with the bog water level, but also with the characteristics of the vegetation cover [6]. Low and tall ryams in the upper 60 cm layer have a lower temperature than the ridge at RHC, where the wood layer is very tenuous. Ryams due to the dense vegetation cover receive less solar radiation, and in conjunction with a low level of bog waters become colder than ridge at RHC.

In winter, the ridge at RHC is the coolest site, because the thickness of the snow cover is less than at ryams, where shrubs and trees linger the snow. A hollow at RHC and an open fen in winter even with small snow depth is warmer than a ridge at RHC, as high level of bog water prevents the rapid cooling of the peat deposit. These features affect the time of occurrence of minimum temperatures. In most areas the minimum temperatures on the surface are observed in February (-0.8, -1.7, -2.4 and -2.5 °C for a hollow at RHC, an open fen, low and tall ryam, respectively), and on ridge at RHC the minimum temperature (-5.2 °C) is observed in January (figure 1).

During the study period, the maximum average daily temperature was noted in the top 20 cm layer in 2012 due to the anomaly hot June. The soil temperature ranged from 23.5 (ridge at RHC) to 24.9 °C (low ryam) on the surface and from 16.2 (tall ryam) to 19.5 °C (open fen) at the depth of 20 cm. Deeper than 30 cm the maximum temperature of the peat deposits ranged between 12.6 (low ryam) to 16.9 °C (open fen) at 40 cm and 5.5 (low ryam) to 7.9 °C (open fen) at 240 cm respectively. The maximum temperature of the upper layers (0-60 cm) strongly influenced the air temperature and the features of the peat deposits (peat density, water content, vegetation cover, etc.).

The minimum average daily temperature was noted in the upper 20 layer and it raised from -9.5 to 2.3 °C from the surface to 20 cm, respectively. This distribution was observed due to several factors: formation of a stable snow cover, snow depth and bog water level. For example, in 2015, due to severe frosts in early November, the temperature in the layer 0-20 cm at tall ryam was lower by 3 °C than in the same period in 2012. The greatest frost penetration was observed at tall ryam in 2012, when the soil temperature at 60 cm depth was -0.2 °C as the height of the snow during the winter varied from 0.9 cm in October 2012 to 23.5 cm in March 2013. The snow depth in 2015 varied from 2.8 cm in October 2015 to 96.2 cm in March 2016 and even when strong frosts in December 2015 did not allow to freeze peat deposit for more than 30 cm depth.

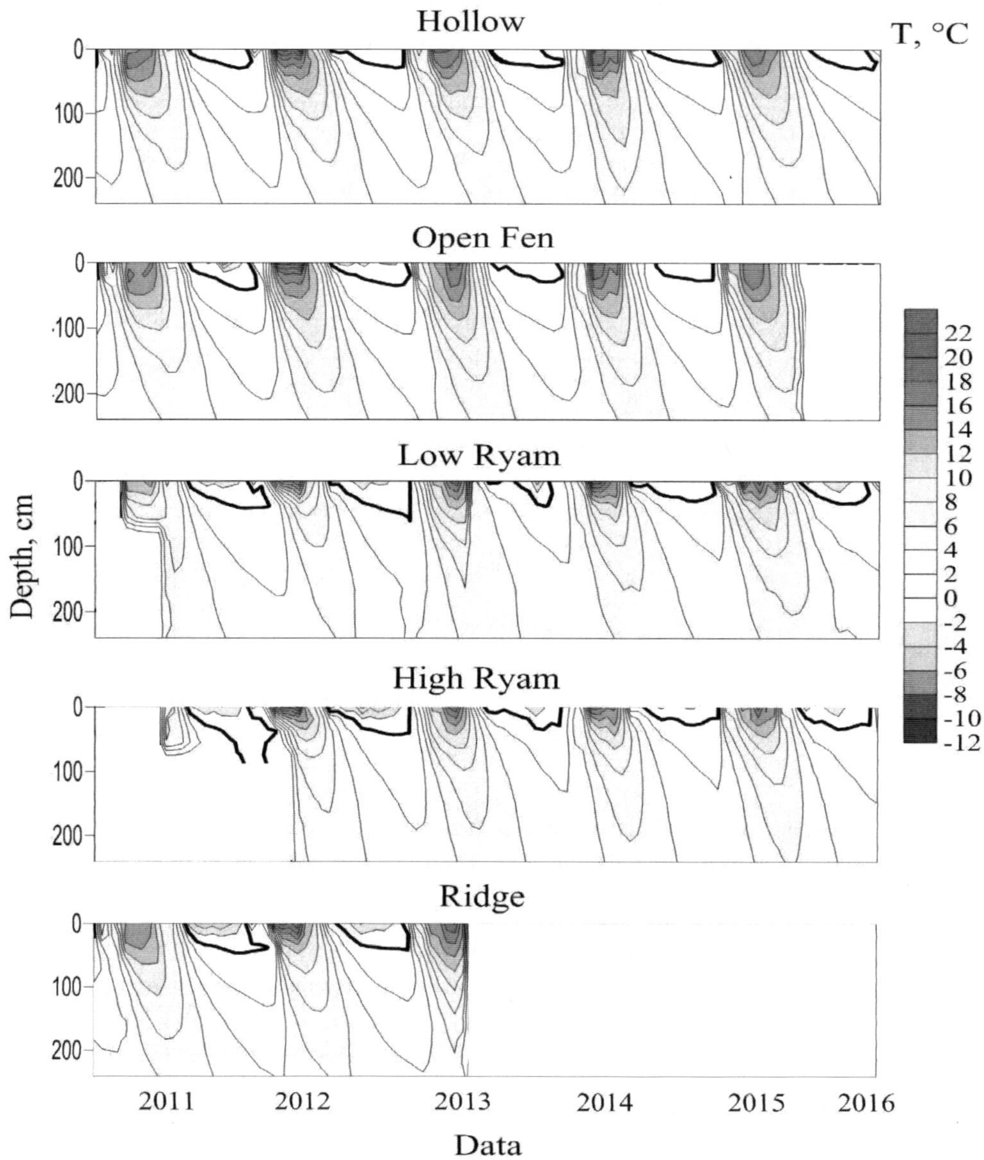


**Figure 1.** Distribution of daily average soil temperature in the studied peat deposits.

The temperature gradient in the soil has the highest variations in the layer 0-2 cm. Its value changes from -2.0 °C/cm in February 2012 to 1.1 °C/cm in October 2014 at the open fen, and from -1.1 °C/cm in October 2012 to 0.8 °C/cm in July 2012 at tall ryam. In the warm period the greatest fluctuations in the soil temperature gradient was observed in the layer 2-60 cm. In July they varied in the range from 0.05 °C/cm in open areas (open fen, hollow at RHC) to 0.33-0.36 °C/cm in the forested bogs (tall and low ryam). The value of the gradient is decreasing with depth increase. At 240 cm depth the value of gradient at all sites over the entire observation period does not exceed 0.1 °C/cm. In the cold period, the greatest fluctuations of temperature gradients in the soil were observed in the layer 2-40 cm. In February, they varied from -0.01 °C/cm in open areas (open fen, hollow at RHC) to 0.15-0.21 °C/cm for forested bogs (ridge at RHC, tall and low ryams).

During the propagation of thermal wave from the surface into the soil there is a lag of this wave. figure 2 shows the lag time of occurrence of maximum daily average temperature at different depths for 2011, 2013 and 2015. Timing of the maximum temperature depends not only on the type of site (open, forested), but also on weather conditions of a particular year. The lag of the daily averaged temperature wave in the layer 0-15 cm is almost absent. In some cases, the maximum temperature in the layer of 0-5 cm can be recorded later than in the underlying layer. The greatest variation in the lag of the temperature wave is observed at depths of 20-80 cm and varies from 0 to 50 days. With the depth increase there is a slowdown in the propagation of a thermal wave and at the depth of 240 cm the maximum is reached on average after 100 days. On the surface, the average is 24-27 °C and it is 5-8 °C at the depth of 240 cm depending on site.

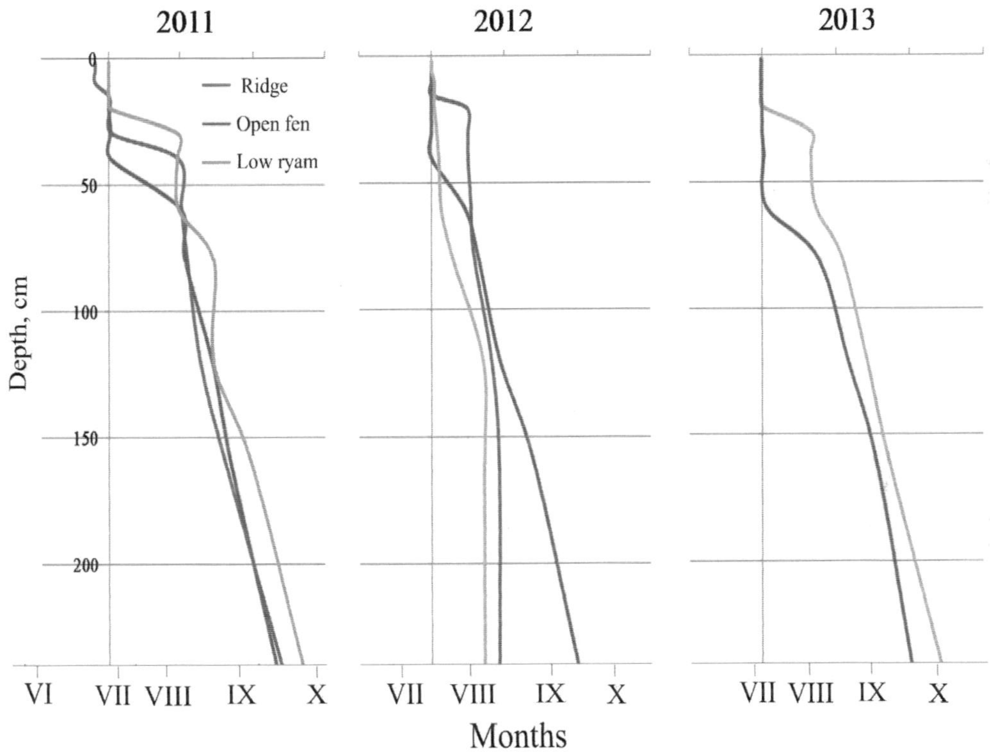


**Figure 2.** Time of occurrence of maximum daily average soil temperature at different depths for ridge at RHC, open fen and low ryam.

In the cold period, the particular importance in the penetration of the temperature wave and the delay of occurrence of minima with depth will play year-round warm lower part of the peat deposit. The layer below 60 cm act as a "battery" of heat, which slows down the process when the soil is freezing. The snow cover reduces the amplitude of temperature fluctuations on the surface and influences the maximum depth of penetration of negative temperatures into the peat deposit [5].

The minimum temperature in the layer 0-30 cm is observed at the beginning of the cold period when the snow cover is minimal or absent. Therefore, in this layer the lag of the temperature wave is not large and ranges from several days to weeks. Increasing the height of snow cover leads to a slowing of thermal wave and a significant increase in the delay of the minima at depths below 60 cm.

The formation of a seasonally frozen layer starts in October-November at all sites. In these months in the South Taiga zone of Western Siberia there is a decrease in the air temperature and precipitation in the form of snow, which in the subsequent period, forms a steady snow cover [5]. A further lowering of air temperature leads to the formation of the seasonally frozen layer at the depths of 40-60 cm in February-March. In the same period, the maximum freezing depth of peat deposits is reached (table 1).

**Table 1.** Maximum depth of freezing of the peat deposit (cm) and their dates.

| Sites | Data | | | | |
|---|---|---|---|---|---|
| | 2012 | 2013 | 2014 | 2015 | 2016 |
| Hollow RHC | 16 March | 14 April | 3 April | 16 April | 9 April |
| | 33.1 | 39.1 | 28.4 | 27.1 | 33.1 |
| Open fen | 18 April | 21 April | 4 April | 15 April | – |
| | 44.1 | 36.1 | 33.9 | 32.4 | – |
| Low ryam | 18 March | 15 April | 25 February | 10 March | 10 March |
| | 40.2 | 51.8 | 38.6 | 31.4 | 36.2 |
| Tall ryam | 12 April | 28 March | 3 March | 19 April | 2 February |
| | 61.4 | 42.5 | 35.0 | 32.6 | 35.1 |
| Ridge RHC | 13 March | 9 April | – | – | – |
| | 48.7 | 41.8 | – | – | – |

The maximum duration of the existence of seasonally frozen layer was observed in the layer of 0-15 cm and the average is 140-160 days, the lowest duration was at the depths below 40 cm and the lowest average was of 80 days. Peat soil freezing proceeds with a rate of 0.2-0.3 cm/day on average from the surface up to the maximum depth. The maximum recorded rate of freezing of 0.51 cm/day was observed in 2012 at tall ryam.

Thus if we split the rate of freezing into two parts, freezing is more intense in the layer 0-20 cm than in the layer from 20 cm to the maximum depth of frost penetration. The rate of freezing from the surface to 20 cm is 0.6-0.7 cm/day on average. But the maximum freezing rate may vary considerably, for example, in 2011 it reached 1.43 cm/day on the ridge, and in 2015 it was 1.67 cm/day in tall ryam. The freezing rate is higher in forested bogs than in open areas. The average freezing rate is 0.21 cm/day in the layer below 20 cm. The maximum rate (0.81 cm/day) was observed in 2014.

The degradation of the seasonally frozen layer starts both from the top and from the bottom. This contributes to a relatively warm ice-free peat deposit, which is below the seasonally frozen layer. Since mid-to late winter the snow depth is significant, the effect of the negative temperatures of the air for maintaining low temperatures of the seasonally frozen layer is weakened and the process of thawing peat thickness starts from bottom. However, the rate of thawing of seasonally frozen layer is very slow from the bottom at the initial stage and its destruction proceeds faster from the top. Thus, the most significant degradation of the seasonally frozen layer occurs from the top. To estimate the rate of thawing, we use the period of thawing start, which coincides with the stable transition of soil temperature above 0 °C. The rate of peat thawing from the top varies from 1 to 5 cm/day to 30-35 cm/day. In addition to the thawing rate increase from the top, the thawing rate from the bottom also slightly increases and varies from 0.3 to 0.5 cm/day.

## 4. Conclusions

The sites with high water level (hollow at RHC, open fen) warm up better in the summer period and have a higher temperature on the surface in comparison with forested bog sites with lowered water level (ridge at RHC, high and tall ryam). In the cold period, open bogs remain warmer than the forested sites. Warming of the peat deposits at depths of 2 m stops in October due to large lag in the heat wave propagation. The maximum depth of freezing was observed in the tall ryam (61.4 cm) in January 2012, and the greatest duration of existence of the seasonally frozen layer was 195 days.

The rate of soil freezing from the surface to the maximum depth of freezing varies from 0.2 to 0.3 cm/day on average. The degradation of the seasonally frozen layer proceeds simultaneously both from the top and from the bottom. The rate of degradation of the seasonally frozen layer on top exceeds significantly the rate of freezing, and it ranges from 1 to 5 cm/day to the complete disappearance of the frozen layer. The rate of degradation of the bottom is smaller and it varies from 0.3 to 0.5 cm/day.

## References

[1] Alekseeva M N, Preis Y I and Dyukarev E A 2015 *Bulletin of the Tomsk Polytechnic University, Geo Assets Engineering* **32(4)** 143–146
[2] Golovatskaya E A, Dyukarev E A, Ippolitov I I and Kabanov M V 2008 *Reports of the Russian Academy of Sciences* **4** 1–4
[3] Golovatskaya E A and Dyukarev E A 2011 *Russian Meteorology and Hydrology* **36(6)** 84–93
[4] Golovatskaya E A and Dyukarev E A 2012 *Eurasian Soil Science* **45(6)** 588–597
[5] Dyukarev E A 2015 *Earth's Cryosphere* **19(3)** 45–51
[6] Dyukarev E A, Golovatskaya E A, Duchkov A D and Kazantsev S A 2009 *Russian Geology and Geophysics* **50(6)** 579–586
[7] Kiselev M V, Voropai N N and Dyukarev E A 2016 *Reports of higher educational institutions. Physics* **7-2** 93–97
[8] Kurakov S A 2012 *Sensors and systems* **4** 29–32
[9] *The second assessment report of Roshydromet on climate change and their consequences on the territory of the Russian Federation* 2014 (Federal service for Hydrometeorology and environmental monitoring) p 61
[10] Dyukarev E A and Golovatskaya E A 2013 *Geography and Natural Resources* **1** 65–71
[11] Ciais P, Sabine C, Bala G, L Bopp, Brovkin V, Canadell J, Chhabra A, DeFries R, Galloway J, Heimann M, Jones C, Quéré Le C, Myneni R B, Piao S and Thornton P, 2013 *Carbon and Other Biogeochemical Cycles. In: Climate Change 2013: The Physical Science Basis. Contribution of Working Group I to the Fifth Assessment Report of the Intergovernmental Panel on Climate Change* ed Stocker T F, Qin D, Plattner G-K, Tignor M, Allen S K, Boschung J, Nauels A, Xia Y, Bex V and Midgley (Cambridge University Press, Cambridge, United Kingdom and New York, NY, USA) p 570
[12] Romanov V V 1961 *Hydrophysics of bogs* (Leningrad: Gidrometeoizdat) p 359

# Plant component features of forest-bog ecotones of eutrophic paludification in the south of boreal forest zone of West Siberia

**N V Klimova*, N A Chernova and N N Pologova**

Institute of Monitoring of Climatic and Ecological Systems Siberian Branch of the Russian Academy of Sciences, 10/3, Academichesky ave., Tomsk, 634055, Russia

*Email: klimnin@sibmail.com

**Abstract**. Paludified forests formed in transitional forest-bog zone aren't studied enough, inspite of its high expected diversity and large areas in the south of boreal forest zone of West Siberia. In this article wet birch (*Betula pubescens*) forests of forest-bog ecotones of eutrophic paludification are investigated on Vasyugan plain with nutrient-rich calcareous clays as soil-forming rocks. Species diversity and ecocoenotic structure of these phytocoenoses are discussed. They correlated with wetness and nutrient-availability of habitats evaluated with indicator values of plants. The participation of hydrophylous species is increasing as wetness of habitats increasing in the forest-to-bog direction like in mesotrophic paludification series. However the number of species is higher in the phytocoenoses of eutrophic paludification. The share of species required to nutrient availability is also higher, both in number and in abundance. A lot of these species are usual for eutrophic boreal forested swamps with groundwater input and absent in forests of mesotrophic paludification. Accordingly the nutrient-availability of habitats is also higher. All these features we connect with birch to be a forest forming species instead of dark-coniferous and with the influence of nutrient-rich parent rocks, which is evident in forest-bog ecotones of Vasyugan plain gradually decreasing together with peat horizon thickening.

## 1. Introduction

In recent decades forest-bog ecotones attract attention as an integral part of bog ecosystems important for its functioning [1]. They make a valuable contribution to the water and carbon cycles [2, 3]. Due to the variety of habitats, often specific, they provide the basis for biological diversity, appeared on different levels [4]. Being vulnerable to environmental changes the forest-bog ecotones have a great potential in the evaluation of current climate changes and also their consequences using space-for-time substitution approach [5]. Of special importance is the study of forest-bog ecotones in West Siberia, where a significant portion of forested area is swamped (paludified forests occupy 25% of the territory in the middle of forest boreal zone and 11% – in the south of forest boreal zone [6]). Although the share of shallow peat forests (with a peat thickness less than 30 cm) is not so high on the whole (7% – in the middle, 4% – in the south [6]), but in some places it strongly increases, so on Vasyugan plain – up to 13–30% of the study area [7]. Despite the fact that a lot of studies of paludified forests have been carried out [2, 3, 7-9], the knowledge about its floristic, phytocoenotic and ecotope specifity is still insufficient and that leads to significant differences in estimates of resources and processes occurring there [2, 10] and complicates the development of an evidence-based approach to economic use and/or protection of these plant communities.

We study paludified forests formed in the forest-bog ecotones on Vasyugan plain. Because of calcareous clays to be parent rocks, the eutrophic paludification develops there in contrast to mesotrophic paludification of forests on loams [11, 12]. Accordingly wet herbaceous forests with low abundance of sphagnum are characteristic for the study area [7, 12, 13]. Most of them have been burned or cut down earlier, therefore birch (*Betula pubescens*) communities, secondary after dark-coniferous forests, are considered. In this article the floristic composition and ecocoenotic structure of these plant communities are discussed to identify the features of plant component of forest-bog ecotone specific for eutrophic paludification.

## 2. Objects and methods

The study area is east part of Vasyugan plain in Ob-Vasyugan interfluve (figure 1, site 1). It is the south of boreal forest (taiga) zone of West Siberia [14]. The climate is cold and continental. The mean January temperature is –20°C, the mean July temperature being 18°C. The duration of the frost-free period is about four months. The annual precipitation is up to 500 mm, it is about 100–200 mm greater than potential evaporation [14]. The soil-forming rocks are calcareous clays [15]. For comparison we use data on mesotrophic paludification of dark-coniferous forests on loams of Chulym plain [16] (figure 1, site 2). We study plant communities in peripheral areas of small oligotrophic bogs on the north-east edge of Great Vasyugan mire. Aspen-birch-mixherbaceous-sedge communities (the most abundant species are *Betula pubescens, Populus tremula, Carex macroura, Aegopodium podagraria, Calamagrostis obtusata, Equisetum pratense, Rubus saxatilis*) – secondary after dark-coniferous forests of southern taiga [17] – are common in uplands [13], inhabiting dark-humus organo-accumulative soils [15].

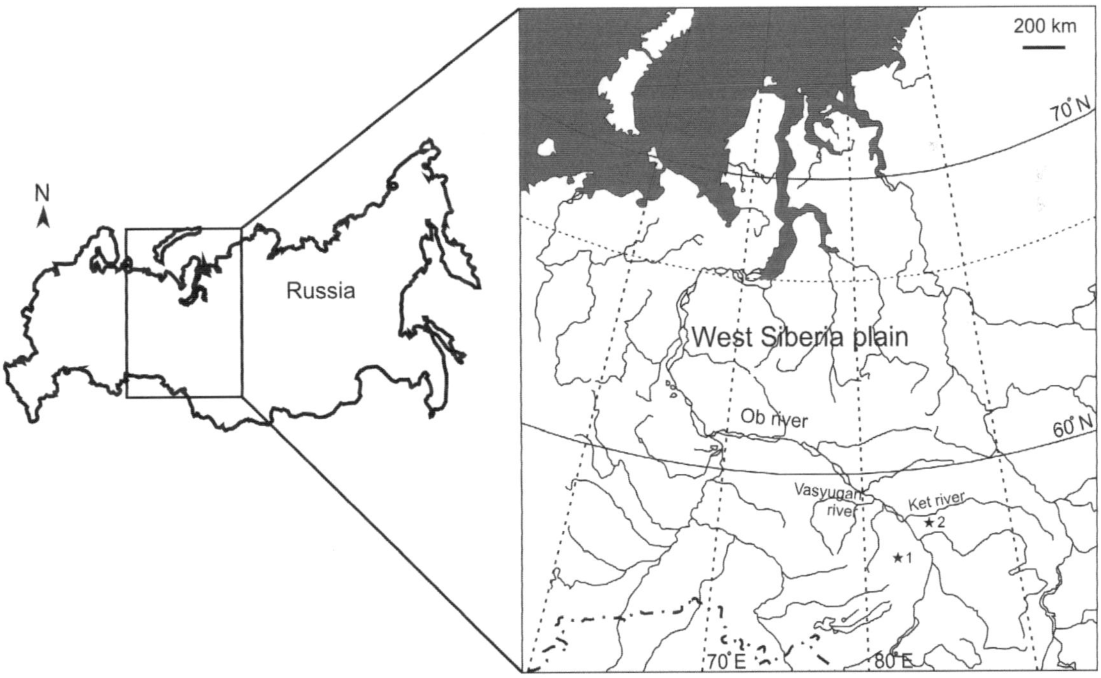


**Figure 1.** Geographical location of the study area. 1 – Vasyugan plain; 2 – Chulym plain.

The pine-dwarf shrubs-sphagnum communities (the most abundant species are *Pinus sylvestris, Ledum palustre, Chamaedaphne calyculata, Sphagnum angustifolium*) occupy bog habitats [13, 18, 19]. We identify two plant associations related to the ecotone parts in transition forest-bog zone given in order to gain influence from the bog: 1) forest part of ecotone with birch-reed grass communities formed on hydromorphic organo-accumulative mucky–dark-humus soils, 2) bog part of ecotone with pine-birch-dwarf shrubs-moss communities formed on periodically flooded peat soils [8]. We investigate 11 bogs. In peripheral area of each of the bogs from 1 to 3-4 study plots (25*25m) were laid out in each part of ecotone and also in adjacent ecosystems for vegetation releves (species list with cover (horizontal projection, %) as abundance measure). On the whole we get 21 vegetation releves in uplands, 25 – in forest part of ecotone, 10 – in bog part of ecotone, 38 – in bogs. Based on this data the floristic diversity of plant associations was analyzed. We use the means of number of species per plot 25 x 25 m as alpha-diversity measure and species richness of a set of vegetation releves as gamma-diversity measure. Interpolated and extrapolated species richness were obtained in EstimateS program [20] using presence/absence data. Jaccard index (Kj) and Morisita index (Km), which both are similarity indices changing from 0 to 1, were calculated to assess beta-diversity according to recommendations [21] using application PAST 1.87b. We analyze ecological and coenotic structure of plant communities with groups of species suggested by E D Lapshina [18]. In addition the wetness and nutrient availability of habitats in forest-bog ecotones were evaluated with indicator values of plants developed by I A Tsatsenkin et al [22]. The calculations of the habitat statuses of these factors were made using the whole floristic species-list of the phytocoenoses and taking into account indicative importance and abundance of each species by recommendations of A A Zverev [23].

## 3. Results

The species richness of birch-reed grass association (forest part of ecotone) does not differ significantly ($p>0.05$) from that of aspen-birch-mixherbaceous-sedge association (upland), where the maximum number of species is found (figure 2). The species richness of pine-birch-dwarf shrubs-moss association (bog part of ecotone) is a little less than that of upland forests. The minimum of species is in bog. The mean species numbers per plot 25 x 25 m have the same pattern (figure 2, for a single sample; figure 3, A).

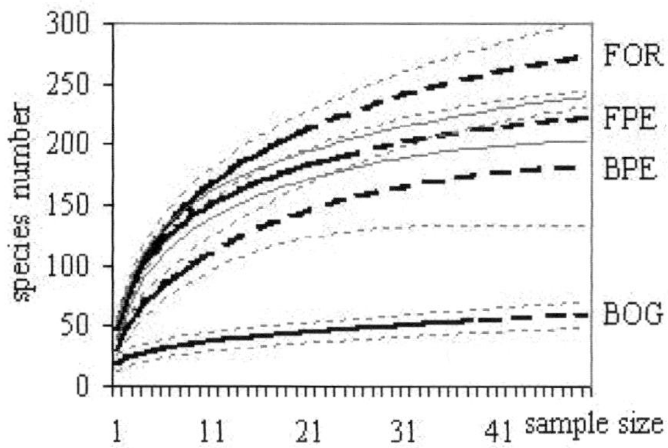


**Figure 2.** Species accumulation curves for the plant communities of upland forests (FOR), transition zone (FPE – forest part of ecotone, BPE – bog part of ecotone) and bogs (BOG). Sample size reflects the number of vegetation releves. Thick solid line is interpolated part and thick dashed line is extrapolated part of the curves. Thin lines limit the confidence area (95 %).

The species composition of upland birch forests and that of bogs have little in common (Kj=0.10, Km=0.05). Pairwise comparison of phytocoenoses across the gradient “forest” – “bog” using presence/absence data (table 1, Kj) shows high rate of species replacement in transition zone (plant communities are less than half-overlapped in species indentity in all pairs). The similarity is higher with species abundance data (table 1, Km). However it is due to shared dominant-species of tree layer in case “forest part of ecotone” – “upland forests” and in lower degree in case “forest and bog parts of ecotone”, because without taking tree species into account the similarity of phytocoenoses is much lower. Such results indicate the higher sensitivity of ground layer species than tree species to the strengthening hydromorphism of habitats under bog influence.

The ecocoenotic structure of upland forests and that of bogs are rather simple. In forests birch and aspen form a tree layer; forest species, which are mostly mesophytic and mesotrophic, dominate in well-developed herb layer both in number and in abundance (cover) (figure 3). In bogs oligotrophic species have the maximum abundance; the sparse tree layer is formed by the bog form of pine (*Pinus sylvestris f. uliginosa*, hight from 8–10 m to 12–15 m, trunk diameter from 10–15 cm to 20–25 cm). In plant communities of forest part of ecotone tree layer is composed with birch and aspen as in uplands.

**Table 1.** Comparison of plant communities across the gradient "forest" – "bog".

| Pairs of phytocoenoses | FOR–FPE | FPE–BPE | BPE–BOG |
|---|---|---|---|
| Jaccard index (Kj) | 0.45 | 0.43 | 0.36 |
| Morisita index (Km) | 0.55/0.08 | 0.58/0.26 | 0.74/0.80 |

Note. There is value of index for the communities in a whole (before slash) and without taking into account tree species (after slash). Abbreviations of communities are the same as in figure 2.

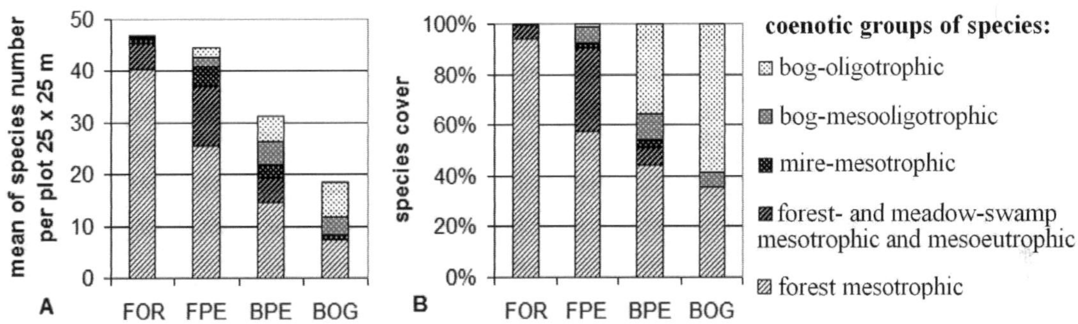


**Figure 3.** Coenotic spectra of plant communities across gradient "forest" – "bog" (abbreviations are the same as in figure 2) in accordance with species number per plot 25 x 25 m (A) and with species cover (B).

However in ground layer species from different coenotic groups co-dominate, all of them are more hydrophilous: meadow-swamp *Calamagrostis purpurea s.l.* (20 % of cover), forest *Equisetum sylvaticum* (10 %) and bog mesooligotrophic *Carex globularis* (10 %). Besides the occurrence of forest- and meadow-swamp species is higher than in uplands, that is reflected both in larger species number per plot 25 x 25 m and also in greater abundance of this coenotic group. Forest- and meadow-swamp species as well as forest species are mesotrophic. The presence of those and other in ground layer of forest part of ecotone is well coordinated with the comparatively high nutrient-availability of both habitats (table 2, FOR and FPE). In phytocoenoses of bog part of ecotone tree layer is composed with birch and pine. Compared with bog one the pine is better developed – forest form (hight 16–20 m, trunk diameter 25–30 cm). In ground layer oligotrophic species, which bog dominating, have high cover (*Ledum palustre* (10 %), *Chamaedaphne calyculata* (5 %) and *Sphagnum angustifolium* (35 %)). But in contrast to bogs hydromesophytic forest, forest- and meadow-swamp species (as in forest part of ecotone) constantly meet, and also mesooligotrophic bog species meet, which have lower cover and occurrence in bogs or absent there. Accordingly the nutrient availability of bog part of ecotone habitat is higher than in bog one, while the wetness statuses of both – coinside (table 2).

**Table 2.** Wetness and nutrient-availability statuses of habitats across gradient "forest" – "bog".

| | Eutrophic paludification (Vasyugan plain) | | Mesotrophic paludification (Chulym plain) | |
|---|---|---|---|---|
| | Wetness | Nutrient-availability | Wetness | Nutrient-availability |
| FOR | 68,2+1,2 | 9,0+0,3 | 68,8+0,7 | 7,5+0,2 |
| FPE | 72,5+1,0 | 8,4+0,4 | 71,6+1,0 | 7,4+0,2 |
| BPE | 75,3+2,0 | 7,4+0,3 | 73,6+1,1 | 7,2+0,2 |
| BOG | 75,1+1,4 | 7,0+0,2 | | |

Note. There are values of average and standart deviation. Abbreviations are the same as in figure 2.

Therefore, the plant communities of forest-bog ecotones of Vasyugan plain are stand out with its floristic composition. There are species, which are absent in adjacent upland forests and in bogs (or which met there rarely with less abundance). A lot of them are forest- and meadow-swamp and better presented in phytocoenoses of forest part of ecotone, the last are mire mesotrophic and bog mesooligotrophic, growing mostly in bog part of ecotone. On the whole these species put together up to 40% of floristic composition and 20–40% of abundance (cover) of ground layers in plant communities of forest-bog ecotones on Vasyugan plain.

## 4. Discussion

The wet birch (*Betula pubescens*) forests of forest-bog ecotones of Vasyugan plain have the features, which differ them from the relative plant communities of another regions of West Siberia. It is difficult to compare the numbers of species because the published estimates concern the taxa of different levels or different area-size. Thus we could admit, that there are far less number of species found in dark-coniferous forests composing mesotrophic paludification series on Chulym plain (113 plant species in 107 samples), however it is situated further to north, on the boundary between middle and southern taiga [16]. In the southern taiga of West Siberia there is also less number of species showed (for comparison: 122 species for well drained herbaceous dark-coniferous forests and 103 species for wet dark-coniferous forests with sphagnum cover [12]). Relatively high plant species richness in birch (*Betula pubescens*) forests of Vasyugan plain, which are secondary after dark-coniferous, may be connected with more sun light penetrating through tree layer and besides with better soil qualities caused by change of leaf fall and litter, that draw habitats together with habitats in aspen-birch forests, replaced dark-coniferous in the southern boundary of boreal forest (taiga) zone, and lead to settling more southern plant species. Thus, there are more such species in secondary birch forest of southern taiga than in dark-coniferous one [17]. Besides, the nutrient-rich soil-forming rocks of Vasyugan plain seem to influence species number of plant communities in the same way – by drawing habitats together with more southern.

As for ecocoenotic structure, wet birch (*Betula pubescens*) forests of eutrophic paludification investigated and wet dark-coniferous forests of mesotrophic paludification [12, 16] have likeness and difference. In both series the number of forest species decreases and the number of hydrophilous bog species increases together with the degree of paludification from upland to bog. Accordingly wetness statuses of correspondent habitats are similar (table 2), while nutrient-availability statuses differ, and also ecological structure of the plant communities is not the same. In the forest part of ecotones of Vasyugan plain in relation with more nutrient availability of ecotopes the participation of mesotrophic species is much higher. For all that a lot of forest- and meadow-swamp species of studied phytocoenoses are usual for eutrophic boreal forested swamps of West Siberia with groundwater input united in order *Calamagrostio purpureae-Piceetalia obovatae* of class *Alnetea glutinosae* of ecologo-floristic classification and described by E D Lapshina [19] – herbs *Filipendula ulmaria, Carex canescens, C. disperma, C. vaginata, C. cespitosa*, mosses *Calliergon cordiifolium, Climacium dendroides, Plagiomnium ellipticum, Sphagnum squarrosum* and others. These species are not met in forests of mesotrophic paludification series. Its presence in the wet birch (*Betula pubescens*) communities of forest part of ecotone we connect with high soil nutrient availability caused by meadow type of humus-accumulation in weakly-acid media turned out on calcareous parent rocks [8]. In the bog part of ecotone the status of nutrient availability is lower than in forest part. That is why the listed upper oligotrophic species which grow in ombrotrophic bogs with moisture input (class *Oxycocco-Sphagnetea* of ecologo-floristic classification [19]) compose the ground layer of paludified forests together with mesooligotrophic species (*Carex globularis, S. russowii*) which usually more abundant in the peripheral areas of oligotrophic bogs in plant communities transitional between paludified *Pinus sylvestris* forests of class *Vaccinio-Piceetea* on the one hand and forested with *Pinus sylvestris f. uliginosa* bogs of class *Oxycocco-Sphagnetea* on the other hand. The presence of these species draw the pine-birch forests of Vasyugan plain (eutrophic paludification) together with corresponded forests of Chulym plain (mesotrophic paludification), where the same species are dominating. However the status of nutrient availability of the first is a little higher yet. It is displayed in constant presence of forest- and meadow-swamp species (although they are not so abundant as in phytocoenoses of forest part of ecotone). Such floristic composition and ecocoenotic structure of pine-birch communities of Vasyugan plain we connect with soil nutrient availability, higher than in oligotrophic bogs. The accumulation of eutrophic peat (sedge-tree), turned out there [8], confirm our conclusion.

## 5. Conclusion

The plant component features of forest-bog ecotones of Vasyugan plain are caused largely by birch to be a forest forming species instead of dark-coniferous and by biogeochemical background of the clayey calcareous parent rocks. It determines eutrophic way of forest paludification on the study area. The influence of high nutrient availability of ecotopes is evident as in upland forests and also in paludified communities across the gradient "forest"–"bog" up to habitats where the process of forming nutrient-rich litter is replaced with the process of peat accumulation. High level of soil nutrient availability cause abnormally high participation (and even dominating) of mesotrophic species in phytocoenoses of forest-bog ecotones. A lot of these species usually grow in eutrophic boreal forested swamps with groundwater input and absent in forests forming mesotrophic paludification series. Besides high nutrient availability cause high species number in the plant communities of uplands and of forest-bog transition zone. That is why the ecotone effect is revealed only from bog.

## Acknowledgements

The research was conducted in the framework of the basic project of Siberian Branch of the Russian Academy of Sciences VIII.77.1.3.

## References

1. Howie S A and Tromp-van Meerveld I 2011 *Wetlands* **31** 613–22
   Paradis E, Rochefort L and Langlois M 2015 *Plant Ecology* **216** 999–1018
   Minayeva T Yu, Bragg O M and Sirin A A 2017 *Mires and Peat* **19(01)** 1–36
2. Efremov S P, Efremova T T, Pimenov A V and Sedel'nikova T S 2012 *Diversity and Dynamics of Forest Ecosystems of Russia* (Moscow: Partnership of scientific publications KMK) **1** pp 190–226
3. Glagolev M V, Ilyasov D V, Terent'eva I E, Sabrekov A F, Krasnov O A and Maksyutov S S 2017 *Atmospheric and Oceanic Optics* **30(4)** 301–9
4. Pystina T N, Zheleznova G V, Kolesnikova A A, Kosolapov D A and Poletaeva I I 2010 *Forest Science* **1** 3–11
   Minayeva T Yu and Sirin A A 2012 *Biology Bulletin Reviews* **2(2)** 164–75
   Mieczan T, Tarkowska-Kukuryk M and Bielanska-Grajner I 2012 *Annales de Limnologie, International Journal of Limnology* **48** 323–36
   Howie S A, van Meerveld H J and Hebda R J 2016 *Mires and Peat* **18(23)** 1–13
5. Hartshorn A S, Southard R J and Bledsoe C S 2003 *Soil Science Society of America Journal* **67** 1572–81
   Payne R J, Creevy A, Malysheva E, Ratcliffe J, Andersen R, Tsyganov A N, Rowson J G, Marcisz K, Zielinska M, Lamentowicz M, Lapshina E D and Mazei Y 2016 *Soil Biology and Biochemistry* **98** 18–21
   Ratcliffe J L, Creevy A, Andersen R, Zarov E, Gaffney P P J, Taggart M A, Mazei Y, Tsyganov A N, Rowson J G, Lapshina E D and Payne R J 2017 *Science of the Total Environment* **607–608** 816–28
6. Glebov F Z and Uskova L M 1984 *Botanical Journal* **69(12)** 1634–40
7. Vasiliev S V and Peregon A M 2003 *Tomsk State University Journal* **7** 38–48
8. Pologova N N and Dyukarev A G 2012 *Research of Natural and Climatic Processes on the Great Vasyugan Mire* (Novosibirsk: Publishing House of SB RAS) pp 214–33

IOP Conf. Series: Earth and Environmental Science **138** (2018) 012007 doi:10.1088/1755-1315/138/1/012007

9. Belenko A A and Sinyutkina A A 2016 *Izvestia of Irkutsk State University* **16** 3–16
10. Bohn T J, Melton J R, Ito A, Kleinen T , Spahni R, Stocker B D, Zhang B, Zhu X, Schroeder R, Glagolev M V, Maksyutov S, , Brovkin V, Chen G, Denisov S N, Eliseev A V, Gallego-Sala A, McDonald K C, Rawlins M A, Riley W J, Subin Z M, Tian H, Zhuang Q and Kaplan J O 2015 *Biogeosciences* **12(11)** 3321–49
    Terent'eva I E, Sabrekov A F, Glagolev M V, Lapshina E D, Smolentsev B A and Maksyutov Sh Sh 2017 *Water resources* **44(2)** 297–307
11. Lvov Yu A 1976 *Theory and Practice of Forested Swamp Science and Hydroforestsoilimprovement* (Krasnoyarsk) pp 36–44
12. Gorozhankina S M and Konstantinov V D 1978 *Geography of Taiga of West Siberia* (Novosibirsk: Nauka)
13. Chramov A A and Valutsky V I 1977 *Forest and Swamp Phytocoenoses of East Vasyuganie* (Novosibirsk: Nauka)
14. Kremenetsky K V, Velichko A A, Borisova O K, MacDonald G M, Smith L C, Frey K E and Orlova L A 2003 *Quaternary Science Reviews* **22** 703–23
15. Dyukarev A G and Pologova N N 2011 *Eurasian Soil Science* **44(5)** 480–92
16. Pologova N N, Chernova N A, Klimova N V and Dyukarev A G 2013 *Forest Science* **3** 38–48
17. Lashchinsky N N and Korolyuk A Yu 2015 *Vegetation of Russia* **26** 85–107
18. Lapshina E D 2004 *Flora of Swamps in the South-East of West Siberia* (Tomsk: Tomsk University Press)
19. Lapshina E D 2010 *Vegetation of Swamps in the South-East of West Siberia* (Novosibirsk)
20. Colwell R K 2013 *Estimates: Statistical estimation of species richness and shared species from samples Version 9* User's Guide and application published at: http://purl.oclc.org/estimates
21. Magurran A E 2004 *Measuring Biological Diversity* (Blackwell Publishing)
22. Tsatsenkin I A, Savchenko I V and Dmitrieva S I 1978 *Methodical Instructions for Ecological Appreciation of Feed Grassland of Tundra and Forest Zones of Siberia and Far East by Plant Cover* (Moscow)
23. Zverev A A and Babeshina L G 2009 *Tomsk State University Journal* **325** 167–73

# Soil temperature of peatland landscapes as a factor in the development of exogenous processes of biogenic relief formation in engineering development of territory

**S Korkin[1], O Talyneva[1], E Kail[2]**

[1] Scientific-research laboratory of geoecology of Nizhnevartovsk State University, Nizhnevartovsk, Russia
[2]Research and Design Institute for Gas Processing, Krasnodar, Russia

Email: egf_nv@mail.ru, Olgapivk@yandex.ru, leka_kail@mail.ru

**Abstract**. In the presented work we consider mire landscapes in the context of temperature monitoring. The mire landscapes in engineering development of the territory are very sensitive to anthropogenic impact that leads to a change in surface conditions, changes in natural succession and, as a rule, to changes in soil temperature and properties, which in turn may develop a complex of hostile geodynamic processes. For this study we used recording systems for field measurement of peat and subsoil temperatures. The measurements were made in two key areas: the territory of the north-taiga landscapes of Western Siberia (the Siberian Ridges), and the territory of the middle-taiga landscapes of Western Siberia (the Ob middle-river lowland). The paper analyses the data obtained from five observation sites (3, 5, 5a, 6 and 8) referred to hydromorphic landscapes. For the territory of the Siberian Ridges the 5-year average soil temperature was 3°C. For the Ob middle-river lowland the 6-year average soil temperature was 4.2°C. The annual soil temperature in the period 2015-2016 for Site 5a (man-disturbed area) was 8.3°C at all depths, which is 3.8°C higher than in a natural bog (Site 5 was a control area).

## 1. Introduction

The peatland landscapes occupy a fairly large part (more than 35%) of the territory in the middle taiga zone of Western Siberia; the wetland cover of an individual landscape province reaches 70%. Mires support the stability of the environment and they contribute to the thermal and water balance formation [2]. The study area is subjected to intense engineering development, in connection with which the natural conditions are disturbed. Since mire landscapes are very susceptible to anthropogenic load, this may lead to adverse consequences for humans. The engineering development leads to a change in surface conditions, to a change in natural succession and, as a rule, to a change in the temperatures and properties of the soil, which in turn may develop a complex of geodynamic processes. The team of authors in monograph of 2010 "Cryogenic Geosystems" noted the dependence of grounds temperatures on landscapes and vegetation characteristics [10]. A number of scientists [1, 11, 12, 14] in their studies indicate that the response of the upper horizons of the lithosphere to modern climate changes strongly depends on the landscape-geological conditions. They also note the existence of a certain correspondence between the parameters of frozen rocks and the hierarchical level of landscapes [7]. The mire landscape as any dynamic geosystem has its own cycles, stages of development and limit of stability. The temperature indexes of grounds have the greatest influence on the manifestations of exogenous processes for bog landscapes within the framework of biogenic

IOP Conf. Series: Earth and Environmental Science **138** (2018) 012008 doi:10.1088/1755-1315/138/1/012008

morphogenesis. The soil temperature is a key factor controlling many biotic and abiotic processes occurring in the soils: growth and productivity of vegetation, decomposition and mineralization of soil organic matter, emission of greenhouse gases [22, 23, 24], the release of dissolved organic carbon [25]. The north and middle taiga zones of Western Siberia are located within sporadic permafrost. A very important indicator, that we have obtained, is the dynamic component of soil temperature within the southern permafrost zone for the key sites [8].

## 2. Object and methods

The aim of the study is to monitor the temperature regimes of the north and middle taiga subzones of bog landscapes to reveal the dynamics of biogenic relief formation in the process of engineering development of the territory.

The measurements were made in two key areas: the territory relating to the north-taiga landscapes of Western Siberia (the Siberian Ridges – observation Sites 3 and 8), and the territory relating to the middle-taiga landscapes of Western Siberia (the Ob middle-river lowland – Sites 5, 5a and 6).

The purpose of the research is to determine the temperature regime of the bog landscapes of the West Siberian Plain, which are under conditions of biogenic relief formation. The first key area is located in the subzone of the northern taiga and is under conditions of rare sporadic frozen grounds at different natural conditions. This characteristic allows us to obtain net data indicators for undisturbed landscapes in the conditions of the northern taiga. The choice of locations for arranging temperature observation sites was based on a number of different criteria: position in the relief, lithological composition, microclimate conditions and surface moistening, composition of soil and vegetation cover. The mire landscapes within this key area were presented by Sites 3 and 8. The third observation site started to operate on 22.07.2010. Site 3 is located in the high hummock-ridge bog at Meggen-Neg-Kui [3]. Site 8 started to operate on 12.08.2011. It is located 1.2 km to the north-west from the meteorological station at the permafrost peat on the hummock-ridge bog. The peat core examined at the hummock-ridge bog is of 2 m thickness. The morphometric parameters of the low ridge were the following: height is 70 cm, width is 4 m, length is 10 m. Poor decomposition of thawed peat was observed from the surface down to 50 cm depth. The thickness of the permafrost peat is 20 cm starting from 50 cm deep down to 70 cm deep from the surface of the low ridge.

The second key area was arranged in the subzone of the middle taiga located on the right bank of the Ob middle-river lowland. In connection with the increased man's activity associated with the oil production industry, the observation sites were established in bog landscapes. Site 6 (with temperature sensors located at depths 20, 40, 60, 100, 200 and 300 cm started to operate on 15.11.2010) is situated in natural conditions, Site 5a (depths 20, 40, 60, 100, and 200 cm) and Site 5 (control point) are in the collector corridor of pipelines. The observations at Site 5 were started in 2010, but no data were received. They were restarted in 2015 after coping with technical problems. The air temperature for the second key area was obtained by means of a temperature registering sensor installed at Site 6.

In this paper we used the methods of recording systems designed for in-situ measurement of the soil temperature. The method is described in details in [3, 4, 13]. The researches for the temperature regime features in the peat deposits of oligotrophic bog in the southern taiga of the Western Siberia are presented in the works [16, 17, 18]. A great scientific interest has been drawn to the results of temperature measurements of the air and soil made by the monitoring network deployed during the implementation of the German-Russian project "Kulunda" [15]. The international experience in studying the soil temperature regimes is important for monitoring development [19, 20, 21]. A temperature recorder DS1921G-F5 (Dallas Semiconductors, USA) with a temperature range from -40 °C to + 85 °C with the sensitivity of 0.5 °C was used for measuring the soil temperatures at the depths of 20, 40, 60, and 100 cm. For the depths from 1 m and deeper we used a temperature recorder DS1921Z-F5 with a temperature range from -5 °C to + 26 °C and with a sensitivity of 0.125 °C. To monitor the air temperature we used a temperature recorder DS1921G-F5. The periodicity of temperature recording was 4 hours for all loggers used for that. The analysis of the diurnal course was carried out based on 5-6 measurements. The daily mean and then monthly mean values for the soil temperature were calculated based on the observation results. The temperature recorders had protective capsules excluding the interference with the temperature sensitivity by submerging them in the wells equipped with a D50 mm case polyethylene pipe with a thermal insulation at the wellhead. The data were collected once a year in autumn. The obtained full-scale data on the temperature regime of the peat soils were processed using the Microsoft Office Excel 2007 software.

The presented material analyzes the data for three seasons of observations: a. 2013-2014; b. 2014-2015; c. 2015-2016. We haven't given the detailed analysis of the obtained values for soil and air temperatures from 2010 to 2013 in this work since it is already published [5, 6]. For the average annual temperatures to compare in the period from 2010 to 2013 see table 3 and table 4.

## 3. Results

The lowest air temperature was registered in January -26.3 °C (2014), -22.9 °C (2015), -25.7 °C (2016) at the first key site within the studied period (2013-2016). It corresponds to the average long-term value of -23.7 °C [26]. In winter 2013-2014 the coldest month was February with a temperature of -26.9 °C and not January as it used to be for the study area. The high positive monthly average temperature was obtained for July +17.2 °C [26]. The summer 2015 was characterized by the warmest month of June with an average monthly temperature of 17.4 °C, and July was colder by 1.4 °C (see table 1). In July 2016 the average monthly temperature was at 3.6 °C higher than the long-term average. For the first key area the average annual air temperature was -3.3 °C (2014), -2.8 °C (2015) and -1.3 °C (2016).

**Table 1.** Comparative data on the basic air temperature characteristics for sites of the first key area in three observation seasons.

| Month | 2013-2014 | 2014-2015 | 2015-2016 |
|---|---|---|---|
| **December** | -13.8 | -14.1 | -16.2 |
| **January** | -26.3 | -22.9 | -25.7 |
| **February** | -26.9 | -15.5 | -12.1 |
| **June** | 12.9 | 17.4 | 19.8 |
| **July** | 16.5 | 16.0 | 20.8 |
| **August** | 12.0 | 11.8 | 15.5 |
| **Annual average t °C** | -3.3 | -2.8 | -1.3 |
| **Total of plus t °C** | 1540.8 | 1706.2 | 2131.9 |
| **Total of minus t °C** | 2713.9 | 2744.7 | 2594.0 |

For the second key area within the study period (2013-2016), the air temperature in January was: -22.3 °C (2014), -19.2 °C (2015), -21.9 °C (2016), which corresponds to the average long-term data of -22.5 °C [27]. In the cold season 2013-2014, February was the coldest month with a temperature of -24.6 °C. But the average monthly temperature was warmer by 4.1 °C in January and by 2.4 °C in February in the second key area. June 2015 was the warmest month, same as for the first key area, with an average monthly temperature of 19.2 °C, and July was by 2.3 °C colder (see table 2). The average annual air temperature was -0.6 °C (2014), 0.2 °C (2015), 1.3 °C (2016) for the second key area. The long-term annual air temperature was -2.3 °C averaged from 1998 to 2004 according to the meteorological station Nizhnevartovsk [27]. Analyzing the results obtained in table 1 and in table 2, we come to a conclusion that the average annual temperature has increased for the observation period studied in the work.

The temperature is a key factor in biogenic relief formation. It affects the growth and productivity of vegetation and its decomposition. The analysis stated below applies to the average and average monthly temperatures obtained at the key sites. The results of soil temperature monitoring in the bog show general patterns of the solar radiation supply and consumption.

The minimum temperature was -0.5 °C (27.01.2014) at 20 cm depth on Site 3, and 0° C in the period from 10.12.2014 to 16.05.2015. The maximum temperature was 15.5 °C (08.07.2014) and 16.9 °C (04.08.2015). The average annual temperature varied from 3.3 °C in 2014 to 4.6 °C in 2015.

**Table 2.** Comparative data of the main air temperature parameters for the second key section.

| Month | 2013-2014 | 2014-2015 | 2015-2016 |
|---|---|---|---|
| **December** | -10.9 | -12.8 | -14.5 |
| **January** | -22.3 | -19.2 | -21.9 |
| **February** | -24.6 | -10.9 | -8.6 |
| **June** | 14.9 | 19.2 | 19.2 |
| **July** | 16.4 | 16.9 | 20.0 |
| **August** | 14.5 | 12.9 | 17.4 |
| **Annual average t °C** | -0.6 | 0.3 | 1.3 |
| **Total of plus t °C** | 1907.6 | 2176.5 | 2541.2 |
| **Total of minus t °C** | 2041.4 | 2081.6 | 2064.3 |

The minimum soil temperatures in the annual cycle depends on the temperature of the underlying surface, the transitions during water freezing in the peat, the thickness of the snow cover and the amount of heat accumulated by the peat deposit within the warm period [18]. In the period 2014-2015 zero temperature was registered on November 24, 2014 at 20 cm depth on Site 8 and the transition through 0 °C was continued from November 27, 2014 to April 7, 2015. The positive temperature values were observed from May 24, 2015. The average annual temperature was 2.9 °C. The minimum temperature was -5.5 °C (24-25.01.2015), the maximum temperature was 16.5 °C (04.08.2015).

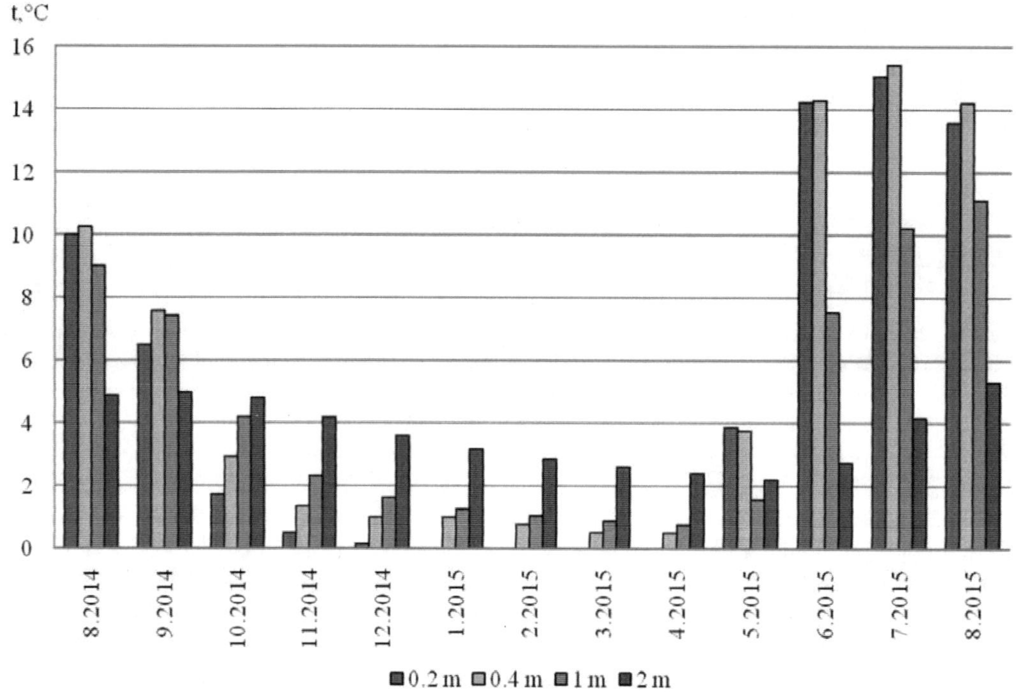


**Figure 1.** Summary graph of the monthly soil temperature at Site 3.

The difference between the soil temperature values obtained at Site 3 and Site 8 in 2015 is due to the landscape factor, since Site 3 is located in a wet hollow and Site 8 is on a ridge. In the period 2015-2016 the zero temperature was observed on November 16, 2015 and the transitions through 0°C were registered from November 20, 2015 to April 1, 2016. The positive temperature values at this depth were observed from June 7, 2016. The average annual temperature was 3.14 °C. The minimum value was 2 °C (26.01-02.02., 05-10.03.2016), the maximum value was 17.5 °C (09.07-10.07.2016). Comparing the totals of positive temperatures we come to a conclusion that the total of negative temperatures has decreased. The total of positive temperatures was 1218.7 °C in 2015 and 1202.4 °C in 2016 and the totals of negative temperatures was 195.0 °C in 2015 and 131.2 °C in 2016.

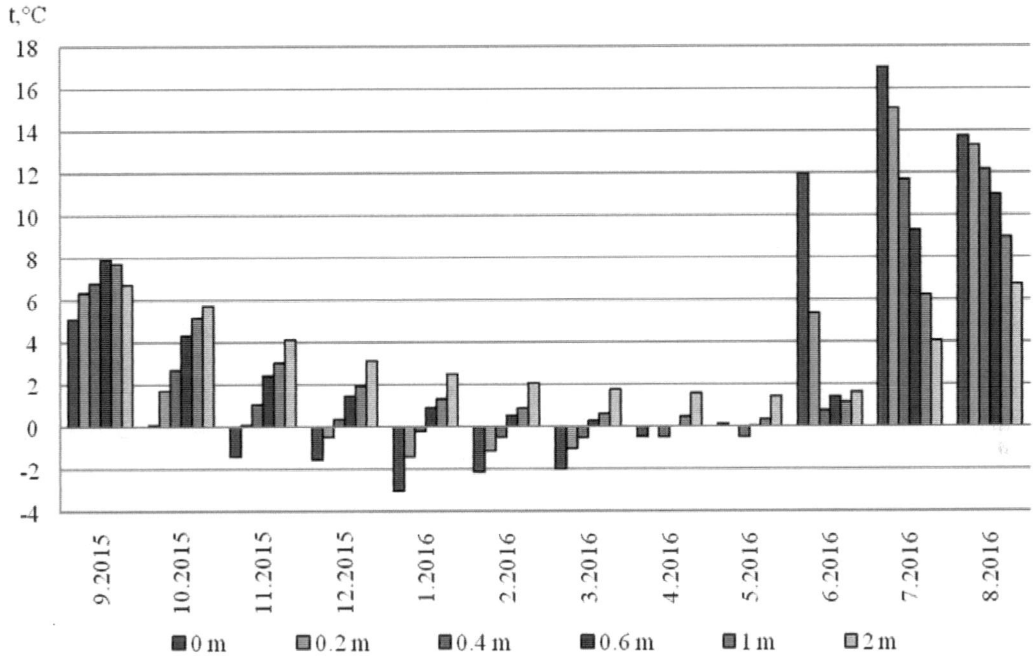


**Figure 2.** Summary graph of monthly soil temperature at Site 8.

The penetration depth of 0 °C temperature into the soil is an important characteristic of the thermal regime in a cold period, which determines the depth of the layer with negative temperatures [18]. In the period 2013-2014 the zero temperature was registered on November 14, 2013 for the depths of 40 and 60 cm at Site 8, and the transition through 0 °C was being registered starting from December 7, 2013 to June 20, 2014. The positive values were noted starting from June 23, 2014. The average annual temperature was 1.2 °C. The minimum value was -1.5 °C (15-28.02.2014), the maximum was 10 °C (August 25, 2014). The total of positive temperatures was 509.7 °C, the total of negative temperatures was -125.1 °C. In the period 2014-2015 the zero temperature was registered on November 30, 2014, and the transition through 0°C was registered in the period from December 25, 2014 to May 26, 2015. The positive values were observed starting from May 30, 2015. The average annual temperature was 2.7 °C. The minimum value was -5.5 °C (24-25.01.2015), the maximum value was 16.5 °C (04.08.2015). The total of positive temperatures was 1044.3 °C, the total of negative temperatures was -99.6 °C. In the period 2015-2016 the zero temperature was registered in December 22, 2015, and the transitions through 0 °C were recorded from January 17, 2016 to June 13, 2016. The positive values started in June 19, 2016. The average annual temperature was 2.8 °C. The minimum value was -1 °C (07-09.03.2016) and the maximum value was 14.5 °C (07.08.2016). The total of positive temperatures was 1001.6 °C, the total of negative temperatures was -74.4 °C. In 2013-2014 the average annual temperature was 1.47 °C at the depth of 60 cm and the total of positive temperatures was 525.9 °C. In 2014-2015 it was 2.98 °C. The minimum value was 0 °C in the period from January 28 to June 30, 2014, the maximum value was 7°C (August 25, 2014). From 05.09.2014 to 04.03.2015, the temperature lowered to 0 °C. The zero temperature lasted until May 25, 2015. After that the temperature raised to positive values and maintained at this point until August 14, 2015. In the period from 01.08 to 14.08.2015 the peak temperature was 11.5 °C. The sum of positive temperatures was 1062.52 °C. In the period 2015-2016 the zero temperature was established from March 21 to May 28, 2016. The average annual temperature was 3.3 °C. The sum of positive temperatures was 1099.7 °C.

In 2013-2014 the average annual temperature for Site 3 was 3.5 °C, in 2014-2015 it was 4.4 °C. At Site 8 the average annual temperature was 1.4 °C in 2013-2014, it was 2.8 °C in 2014-2015 and it was 3.2 °C in 2015-2016. The calculations were made using the average annual temperatures according to the depth values given in table 3.

**Table 3.** Annual average soil temperature at Site 3 and Site 8 of the first key area for the period from 2010 to 2016.

| Observation site | Depth, *m* | Period | | | | | | |
|---|---|---|---|---|---|---|---|---|
| | | 2010-2011 | 2011-2012 | 2012-2013 | 2013-2014 | 2014-2015 | 2015-2016 | Average for 2010-2016 |
| | 0.2 | 3.9 | 4.5 | 3.4 | 3.3 | 4.6 | - | 4.0 |
| | 0.4 | 3.6 | 4.2 | 3.9 | 3.8 | 5.3 | - | 4.2 |
| **3** | 0.6 | 3.6 | 4.2 | 3.2 | - | 4.3 | - | 3.8 |
| | 1.0 | 3.1 | 3.7 | 3.6 | 3.3 | 4.2 | - | 3.6 |
| | 2.0 | - | - | 3.6 | - | 3.6 | - | 3.6 |
| | 0.2 | - | 2.5 | 1.8 | - | 2.9 | 3.1 | 2.6 |
| | 0.4 | - | 1.1 | 1.5 | 1.2 | 2.6 | 2.8 | 1.8 |
| **8** | 0.6 | - | 0.7 | 1.4 | 1.5 | 3.0 | 3.3 | 1.9 |
| | 1.0 | - | 1.1 | - | 1.6 | 2.9 | 3.1 | 2.2 |
| | 2.0 | - | - | - | - | - | 3.4 | 3.4 |

Note: "-" – no date.

No transitions through 0° C were registered at the depth of 1 m. In 2013-2014 the average annual temperature was 3.3 °C at 1 m depth and it was 4.2 °C in 2014-2015. For 2013-2014 the minimum value was 0.5 °C (08.04.2014) and the maximum value was 9.62 °C (08.08.2014). From 01.09.2014 to 16.05.2015 the temperature was 0.63 °C, then from 17.05 to 08.08.2015 the temperature started to rise up to 11.5 °C. In the period from 08.08 to 11.08.2015 the temperature remained at the value of 11.5 °C, after which it started to decline slowly (stick-slip nature of the peak of positive temperature). The maximum value of soil temperature variation was in the period from 08.08 to 11.08.2015 (11.5 °C). The minimum value of temperature variation was 0.6 °C in the period from 08.05 to 16.05.2015. From 01.09.2014 to 28.05.2015 the temperature slowly decreased at the 2 m depth, then from 29.05 to 29.08.2015 the temperature started to rise up to 5.8 °C. The maximum value of the temperature was 5.8 °C in the period from 28.08 to 29.08.2015. The minimum value of the annual temperature variation was 2.13 °C in the interval from 19.05 to 28.05.2015. The average annual temperature was 3.4 °C. In 2013-2014 the average annual temperature was 1.6 °C at the 1 m depth with the total of positive temperatures of 586.3 °C, and it was 2.9 °C in 2014-2015. The minimum value was 0 °C (15.05.-11.06.2014) and the maximum value was 5.5° C (30.08.-02.09.2014). During the period 2014-2015 the average annual temperature was 2.9 °C. The minimum value was 0.3 °C (28.04-24.05.2015), the maximum value was 9.1 °C (12-23.08.2015). The total of positive temperatures was 1015.37 °C. In the period 2015-2016 the average annual temperature was 3.1 °C. The minimum value was 0.3 °C (11-25.05.2016) and the maximum value was 9.3 °C (19.08.-01.09.2016). The total of the positive temperatures was 1047.3 °C.

In the period 2015-2016 the average annual temperature was 3.4 °C at the 2 m depth. The minimum value was 1.3 °C from 11 to 25 May 2016 and the maximum value was 9.3 °C from August 19 to September 1, 2016. The total of the positive temperatures was 1047.3 °C.

For the territory of the first key area in the northern taiga landscapes of Western Siberia, the average temperature for 5 years at the two sites was 3.0 °C, which is 1 °C above the isoline temperature at this latitude given in the paper [28].

The annual average soil temperature at Site 6 had positive values in the observation period (table 4). The highest temperature was 4.8 °C for 2015-2016 and the lowest one was 3.7 °C for 2013-2014. The temperature maintained positive at all depths, without dropping below 0 °C (table 4).

**Table 4.** Annual average soil temperature for Site 6 at the second key area from 2013 to 2016.

| Depth, *m* | 2013-2014 | | | 2014-2015 | | | 2015-2016 | | |
|---|---|---|---|---|---|---|---|---|---|
| | Mean | Max | Min | Mean | Max | Min | Mean | Max | Min |
| **0.2** | 3.5 | 12.7 | 0.0 | - | - | - | 5.7 | 18.2 | -0.2 |
| **0.4** | 4.0 | 11.1 | 0.4 | 5.2 | 13.7 | 0.1 | 5.6 | 15.5 | 0.4 |
| **0.6** | 3.5 | 9.2 | 0.2 | 4.5 | 11.6 | 0.0 | 4.7 | 12.4 | 0.0 |
| **1.0** | 3.8 | 9.2 | 0.5 | 4.7 | 11.1 | 0.5 | 4.9 | 11.7 | 0.6 |
| **2.0** | 3.7 | 5.5 | 2.3 | 3.9 | 6.1 | 2.3 | 4.1 | 6.1 | 2.5 |
| **3.0** | 3.7 | 4.4 | 3.0 | 3.8 | 4.6 | 3.0 | 4.0 | 4.8 | 3.3 |
| **Annual average** | 3.7 | 8.7 | 1.1 | 4.4 | 9.4 | 1.2 | 4.8 | 11.5 | 1.1 |

Note: "-" – no date.

The highest annual average temperatures for all depths of measurements was recorded in the period from 2015 to 2016. The lowest value of average annual temperatures for all depths of measurements was registered in the period from 2013 to 2014. The maximum difference between the values was 2.1 °C at the 20 cm depth. The minimal difference was 0.3 °C at the 2 meters depth. The soil temperature variations are shown in figures 3, 4.

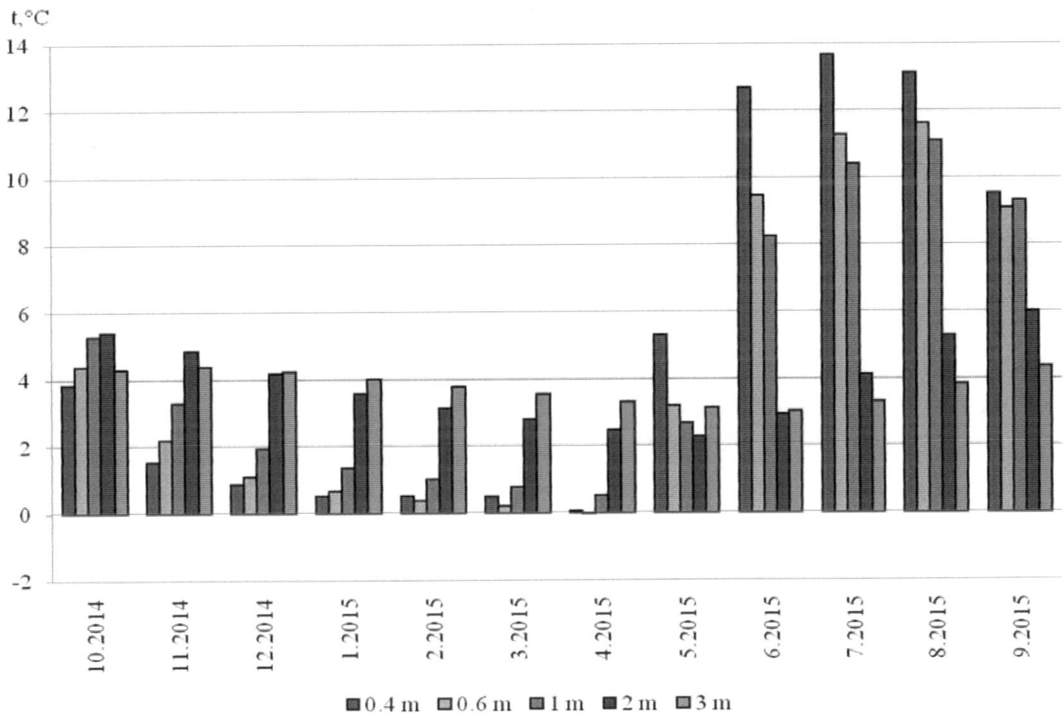


**Figure 3.** Summary graph of the soil temperature at Site 6 for 2014-2015.

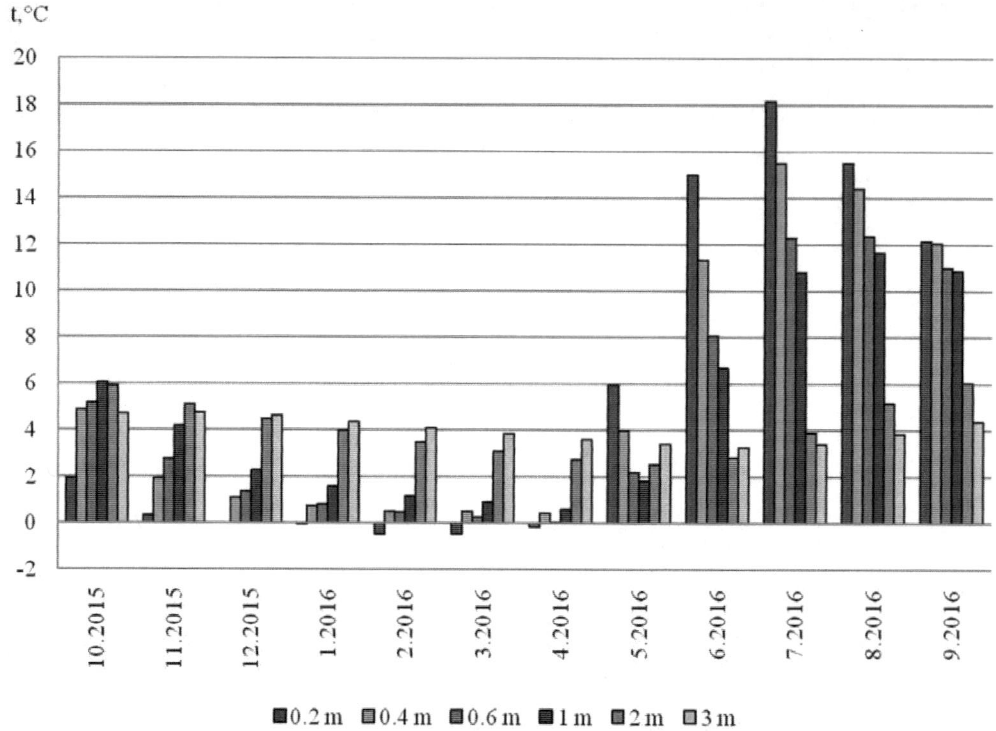


**Figure 4.** Summary plot of the soil temperature at Site 6 for 2015-2016.

The temperature transition through 0 °C was observed on 22.10.12 at the 20 cm depth; it was at the 40 cm depth on 21.01.13 and it was at the 60 cm depth on 01.01.2013. No deeper transition through 0° C was registered. The transition through 0 °C was as well in 2015-2016 at the 20 cm depth on 20.01.16 and no temperature transition was detected below. For the latitude of the Ob river valley the average temperature was 4.2 °C at Site 6 in the period of 6 years.

To study the impact of the pipeline transport on the bog landscapes in 2015, Site 5 was updated and a new site 5a was set up at the man-disturbed area (section of pipeline Samotlor-Nizhnevartovk). The temperature of the oil transported through the pipeline with a diameter of 1220 mm is 17°C. The pipe is laid at the 1 m depth from the surface of the upper hummock-ridge bog. Site 5a is located within the immediate vicinity of the pipeline and Site 5 is located in 20 m from the pipeline. For the period 2015-2016 the annual average temperature at Site 5 was positive with the value of 4.2°C and it is 3.8°C lower than the temperature at Site 5a (table 5).

## 4. Discussions

The bog landscape as a dynamic geosystem has its own cycles and stages of development, as well as a limit of stability. The greatest influence on the manifestations of exogenous processes for bog landscapes is exerted by the soil temperature within the framework of biogenic morphogenesis. For the territory of the first key area, the average temperature was 3 °C at the two sites for the 5 year period. For the Ob middle-river lowland, the average temperature was 4.2 °C at Site 6 for the 6 year period. For the period 2015-2016 the average annual temperature was 8.3 °C at Site 5a (man-disturbed area) at all depths and this value is 3.8 °C higher than in the natural bog (Site 5 – control section). Site 6 is located at the distance of 10 km to the south-west from Site 5. The average annual temperature at Site 5 was less by 0.8 °C than at Site 6. At the 20 cm depth the difference was up to 1.1 °C, at the 40 cm depth it was 2 °C, at the 60 cm depth it was 0.5 °C, at the 1 m depth it was 0.7 °C and at the 2 m it was 0.4 °C.

## 5. Conclusions

It can be concluded that the heating capacity of engineering facilities for pipeline transport is limited within the 25 m protection zone from the pipe line. This fact suggests that the observed higher temperature in the immediate vicinity of the pipeline contributes to the process of restoring the disturbed terrain. The process of biogenic morphogenesis accelerates under temperature increase, and in our case, the peat accumulation process is occurring under restauration of the bog vegetation. In the collective monograph [29], which discloses a single concept of the bog formation process, the average rates of vertical peat accumulation for the northern taiga are 0.37 mm/year and for the middle taiga they are 0.57 mm/year. For a bog massif framing The Lake Samotlor (the second key area) the peat growth rate is 0.32 mm/year, with the peat deposit age of 7700 ± 60 years and the average thickness of 2.5 m [30]. The appearing rate of peat accumulation can't be considered as the same for the whole bog massif, since differences will be observed within the hollows, low ridges and lakes. This rate is a common quantitative indicator. At present, bogging occurs at the expense of the growth of the existing bog systems in broadwise and the emergence of new hearths of bogging in natural conditions occurs rarely. New hearths of bogging arise as a consequence of the man-made activity effect under lack of spillway structures in the places of roads, oil and gas pipelines, impoundment of small rivers and streams. As a result of temperature monitoring of bog landscapes, it is possible to identify the patterns of soil temperature change in natural bog landscapes to a greater extent depending on the influence of meteorological characteristics. For soils under man-disturbed facilities a higher temperature is observed in comparison with the natural bog landscapes, which is a factor of development of heating effects in the process of engineering development of the territory. These results are presented in the database of net soil temperature data of the upper part of the annual heat cycle layer, on the basis of which it is possible to build change forecasts and use data in project-supporting works.

**Acknowledgements**

The work was carried out within the framework of the implementation of the initiative scientific project no. 5.7590.2017/8.9 of the Ministry of Education and Science of Russia.

**References**

[1] Vasiliev A A 2004 *The dynamic of sea shores in the permafrost zone of the Western Sector of the Russian Arctic (on the example of the Kara Sea)* (Tyumen: Trd)

[2] Lopatin K I 2012 *The geoecological foundations for the use of peat bogs and forests of the Middle Ob region: monograph* (Tver: OOO Publishing House Triada)

[3] Kazantsev S A and Duchkov A D 2006 *The high-precision temperature monitoring for geological and geoecological tasks: equipments and the application experience* (Novosibirsk: Siberian State University) pp 25–29

[4] Konstantinov P Ya, Fedorov A N, Michimura T, Iwahana G, Yabuki H, Iijima Y and Costard F 2011 *Cryosphere of the Earth* **XV (1)** 23–32

[5] Korkin S E 2012 *Vestnik of the Tyumen State University* **7** 69–75

[6] Korkin S E and Kail E K 2014 *Izvestiya Samara Scientific Center of the Russian Academy of Sciences* **16 (4)** 1209–12

[7] Korkin S E and Kail E K 2015 *In the world of scientific discoveries* **4 (64)** 330–49

[8] Korkin S E, Kail E K and Mironova N S 2015 *The Samara Scientific Center of the Russian Academy of Sciences* **17 (6)** 104–9

[9] Korkina E A 2015 *The self-regeneration of soils of the Ob middle-river of disturbed by man-made action: monograph* (Nizhnevartovsk: Publishing House of Nizhnevartovsk State University) p 158

[10] Melnikov V P, Khimenkov A N and Brushkov A V 2010 *Cryogenic geosystems: problems of research and modeling* (Novosibirsk: Academic Publishing House "Geo") p 390

IOP Conf. Series: Earth and Environmental Science **138** (2018) 012008 doi:10.1088/1755-1315/138/1/012008


[11] Orekhov P T 2008 *Landscape differentiation of the temperature regime of the soils of the northern taiga of Western Siberia* (Tyumen: Earth Cryosphere Institute SB RAS) pp 252–255

[12] Pavlov A V 2008 *Monitoring of permafrost zone* (Novosibirsk: Academic publishing house "Geo")

[13] Popov Yu A and Borisenko K Yu 2006 *The system of information and recording for the field measurement of soil temperature* In: Theory and practice of assessing the state of the Earth's cryosphere and the forecast of its change: Materials of Intern. Conf. (Tyumen: Tyumen State Oil and Gas Universiy) pp 52–55

[14] Skryabin P N, Skachkov Yu B and Varlamov S P 1999 *Cryosphere of the Earth* **III (3)** 32–40

[15] Belyaev V I and Bondarovich A A 2017 *Vestnik of the Altai State Agrarian University* **3 (149)** 30–7

[16] Dyukarev E A 2012 *Vestnik of Tomsk State University* **365** 201–05

[17] Dyukarev E A, Golovatskaya E A and Duchkov A D 2009 *Geology and Geophysics* **50 (6)** 745–54

[18] Dyukarev E A and Golovatskaya E A 2013 Geography and natural resources **1** pp 65–71

[19] Zang Y, Chen W and Smith S L 2005 *Journ. of Geoph. Res.* **110** D03112

[20] Wisser D, Marchenko S S and Talbot J 2011 *Earth System Dynamics* **2** 121–38

[21] Sheng Y, Smith L C, MacDonald G M, Kremenetski K V, Frey K E, Velichko A A, Lee M, Beilman D W and Dubinin P A 2004 *Global Biogeochemical Cycles* **18** GB3004

[22] Moore T R and Dalva M 1993 *J. Soil Sci.* **44** 651–64

[23] Vompersky S E 1994 *The role of bogs in the carbon cycle (Biogeocenotic features of bogs and their rational use* (Moscow: Nauka) 5–37

[24] Golovatskaya E A and Dyukarev E A 2008 *A report RAS* **418 (4)** 1–4

[25] Prokushkin A S and Guggenberger G 2007 *Meteorology and Hydrology* **6** 93–105

[26] Korkin S E 2014 Monitoring of coastal deformations in the territory of nature park "Siberian Ridges" In: The geoecological monitoring in the territory of the natural park "Siberian Ridges" collective monograph. (Nizhnevartovsk: Publishing House of Nizhnevartovsk State University) pp 100–111

[27] Korkin S E 2008 *Natural hazards of the valley landscapes of the Ob middle-river lowland: monograph* (Nizhnevartovsk: Publishing House of Nizhnevartovsk State University)

[28] Duchkov A.D, Balobaev V T and Devyatkin V N 1995 *Geology and Geophysics* **36** 72–81

[29] Liss O L, Abramova L I and Avetov N A 2001 *The bog systems of Western Siberia and their conservation value* (Tula: Grif and K)

[30] Liss O L, Berezina N A and Kulikova G G 1976 *Natural conditions of Western Siberia* **6** 69–86

IOP Conf. Series: Earth and Environmental Science **138** (2018) 012009 doi:10.1088/1755-1315/138/1/012009

# Testate amoebae analysis in the peat deposits of the swamp Dolgon'koye in the south of Western Siberia and peatland paleohydrology for last 3100 years

**Irina V Kurina and Tatiana A Blyakharchuk**

Laboratory of Monitoring of Forest Ecosystems, Institute of Monitoring of Climatic and Ecological Systems Siberian Branch of the Russian Academy of Sciences, Academichesky ave. 10/3, 634055 Tomsk, Russia

Email: irina.kuryina@yandex.ru

**Abstract**. Our research is devoted to paleohydrological reconstruction in the swamp located in the river valley on the piedmont of the Altai Mountains in the south of Western Siberia. The reconstruction was carried out based on rhizopod analysis for the last 3100 cal yr. A large amount of different testate amoebae was found in the peat. Total 64 testate amoebae taxa were recorded in the peat core with the most abundant being: *Trinema lineare*, *Centropyxis aculeata*, *C. aerophila*, *Euglypha rotunda*, *Cryptodifflugia* sp. Decrease of surface wetness in the swamp are observed 2280, 2140, 1900–600 cal yr BP and increase – in 2700, 2500–1900, 230–215 cal yr BP. The results of our reconstruction of the swamp paleohydrology agrees well with the paleoclimatic data obtained earlier for the central area of the south of Western Siberia Plain. It indicates a high sensitivity of the swamp to climatic changes in the Holocene. The rhizopod analysis proved to be very effective when used for paleohydrology reconstruction in minerotrophic peat.

## 1. Introduction

Our study focuses on the reconstruction of surface wetness dynamics in minerotrophic swamp during its development based on rhizopod analysis data. Rhizopod analysis is a study of fossil testate amoebae remained in the peat. Testate amoebae (Rhizopoda, Testacea) are unicellular organisms, which are widespread in mire ecosystems and often used in paleoecological research as bioindicators for reconstruction of surface wetness dynamics. In most cases, this analysis is applied in researches of ombrotrophic mires, because their surface wetness depends directly on hydroclimatic fluctuations [1-3]. However, evaluating the possibility of using rhizopod analysis for the fens is of special interest. This will allow us to expand the borders of application of the paleoecological reconstructions based on rhizopod analysis for the areas where there are no ombrotrophic bogs. The south of Western Siberia Plain may be considered as one of such areas. According to [4] this area belongs to the eutrophic (phragmites) and brackish bog zone, where peatlands generally occur in valleys or in topographic depressions. For paleoclimatic reconstructions, such natural archives as lake sediments are of primarily use [5-7]. Although one of the works [3] reconstructed the amoeba-based paleohydrology of an ombrotrophic bog located at the border of the southern taiga and the birch-poplar forest steppe, it is important to note that the area under study belongs to the zone of insufficient moisture, so the hydrological conditions in the peatlands of this territory are strongly dependent on the climate

IOP Conf. Series: Earth and Environmental Science **138** (2018) 012009 doi:10.1088/1755-1315/138/1/012009

humidity [4]. Therefore, the peatland deposits are a very promising object for the paleohydrology reconstruction and paleoclimatic studies in this regard.

The purpose of the study is to reconstruct the surface wetness dynamics in the floodplain terrace minerotrophic swamp in the south of Western Siberia during the Holocene. We are posing the following specific questions: 1) What is the structure of the composition and species of the testate amoebae assemblages in the minerotrophic peat? 2) How many tests of amoebae are in minerotrophic peat and is it worth to use the rhizopod analysis for reconstruction of swamp paleohydrology? 3) Are there enough amoeba tests in the peat for quantitative reconstruction of the water table depth (WTD)? 4) What is the sensitivity of the minerotrophic deposits to the paleoenvironmental changes, which may be expressed in presence, frequency, amplitude of fluctuations of reconstructed WTD?

## 2. Study area

The study area belongs to the south of Western Siberia, Kosikhinsky district of the Altai region, to the territory of the third floodplain terrace on the right bank of the Ob' River on a piedmont of Altai mountains [8] (figure 1). The investigated swamp Dolgon'koye lies in the valley of the shallow river Bobrovka in 4 km from the lake Krasilovo (53°12'45'' N, 84°24'26'' E, 198 m a.s.l.). Due to the pristine relief, the swamp shape is elongated from the southwest to the northeast. It is 1600 m long and its maximum width is about 300 m. At the present time, tall birch forest (tree height is 13–15 m) grows on the surface of the swamp with a rich shrub and herb cover. It includes a mixture of forest, meadow, marsh, marsh-swamp and weedy species of plants. The upper 10 cm layer of the peat is highly decomposed, well-structured humus horizon due to the activity of large number of earthworms. The overgrowth of the swamp by the high-stand forest clearly indicates its drainage in the last several decades. In the absence of direct anthropogenic impact on the swamp, this process is possibly caused by the influence of the climate. The peat core was collected in the thickest part of the swamp profile. The thickness of the peat deposit is 287 cm. It is underlaid by 135 cm of laminated organo-mineral deposits. For the peat deposit 10 radiocarbon dates were obtained by liquid–scintillation method in the laboratory of IMCES SB RAS (table 1). The peat deposit is formed by minerotrophic peat with clear layers of different color. Total 32 peat samples for rhizopod analysis were collected with step 1 cm selectively throughout the depth of 210 cm of the peat deposits (from the deeper part of the peat core only small samples were selected due to the very dense structure of the deposits. So they were not sufficient for the rhizopod analysis). The peat sample from the 209-208 cm depth is dated 3090 cal yr BP.

This paper presents preliminary results of the rhizopod analysis using only 32 separated samples from the peat core. In the future, we are planning to continue the research and obtain a continuous series of testate amoeba data for this peat core. To carry out the rhizopod analysis we selected 32 peat samples with the most contrasting results in regards of the obtained peat humification (authors unpublished data). The peat humification was determined by the spectrophotometric method [9].

## 3. Methods

Peat samples (approximate volume is 3 $cm^3$) for testate amoeba analysis were prepared following a modified water-based sieving technique [10]. The samples were mixed with water and thoroughly shaken. The suspension passed through a sieve with mesh opening of 0.355 mm. Then the suspension was concentrated by sedimentation to the volume of 10 ml. The testate amoebae were indentified and counted in the wet mounts using a light microscope (Altami, Russia) with ×200-400 magnification. The quantitative reconstruction of the water table depth during the swamp development is carried out based on testate amoeba data. For these calculations we used the calibration model, which we developed in the other study on some minerotrophic mires in the southern taiga of Western Siberia. The model included 45 moss samples taken from the mire surface. The results of the canonical correspondence analysis confirmed that the relationships between the water table depth and the species structure of testate amoebae assemblages are statistically significant. Therefore, transfer functions could be developed for the water table depth. The transfer function WA (weighted averaging) was applied for the reconstruction. The model performance was achieved with jackknifing (leave-one-out) method: RMSEP – 7.1; $r^2$ – 0.75; Max bias – 11.7; and bootstrapping (999 cycles): RMSEP – 6.8; $r^2$ – 0.79; Max bias – 12.7.

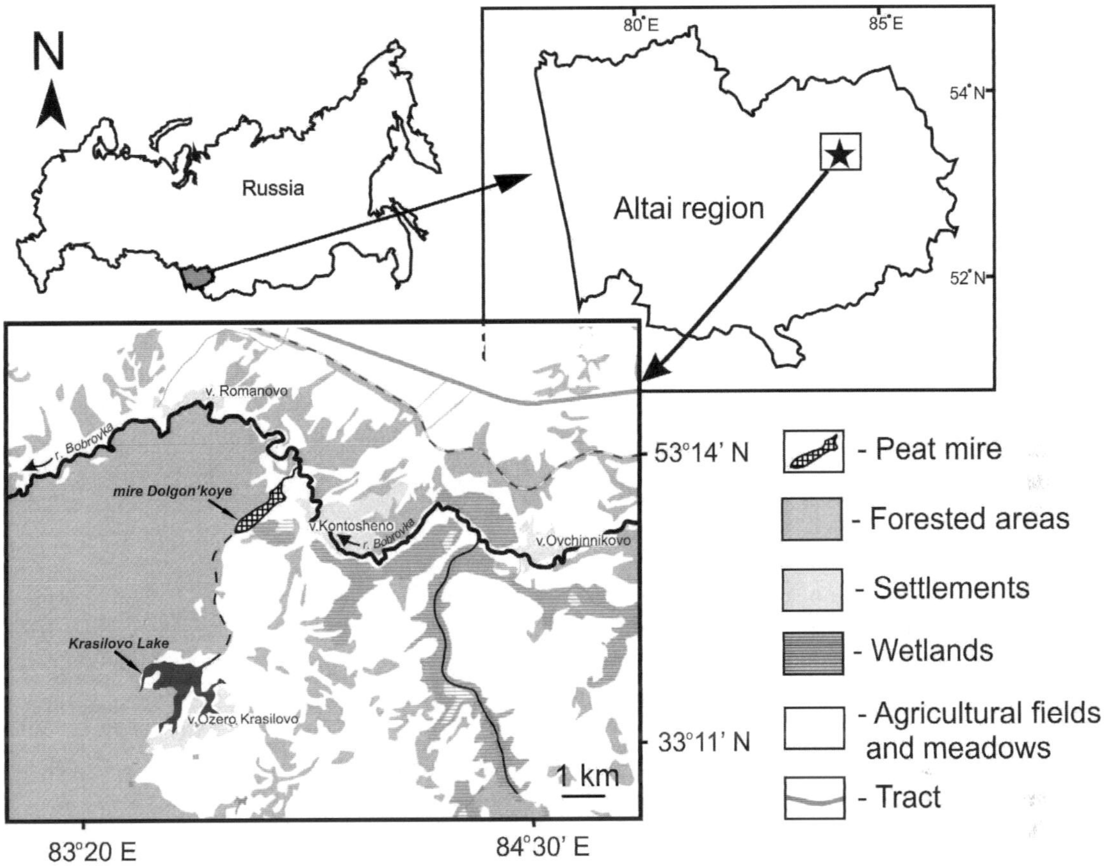


**Figure 1**. Geographical location of the study site (pointed star).

**Table 1**. Radiocarbon dates of the peat core from the swamp Dolgon'koye.

| Lab. No | Dated material | Depth, cm | Age $^{14}$C yr BP | Age cal yr BP (1 σ) | Probability |
|---|---|---|---|---|---|
| IMCES-14C921 | Peat | 36–37 | 572 ± 83 | 616 (584–648) | 0.617 |
| IMCES-14C905 | Peat | 49–50 | 706 ± 111 | 677 (621–733) | 0.675 |
| IMCES-14C547 | Peat | 70–71 | 963 ± 72 | 863 (793–932) | 1.000 |
| IMCES-14C909 | Peat | 80–81 | 1503 ± 68 | 1368 (1319–1417) | 0.745 |
| IMCES-14C926 | Peat | 105–106 | 1929 ± 102 | 1881 (1771–1990) | 0.911 |
| IMCES-14C542 | Peat | 124–125 | 2108 ± 74 | 2073 (1989–2156) | 0.891 |
| IMCES-14C903 | Peat | 132–133 | 2173 ± 111 | 2184 (2051–2317) | 1.000 |
| IMCES-14C546 | Peat | 168–169 | 2398 ± 72 | 2422 (2346–2497) | 0.749 |
| IMCES-14C907 | Peat | 191–192 | 2758 ± 81 | 2853 (2776–2929) | 0.950 |
| IMCES-14C544 | Peat | 208–209 | 2944 ± 63 | 3090 (2999–3180) | 0.980 |

IOP Conf. Series: Earth and Environmental Science **138** (2018) 012009 doi:10.1088/1755-1315/138/1/012009

## 4. Results and discussion

The total 64 testate amoebae taxa were recorded in the peat core with the most abundant being: *Trinema lineare* (average relative abundance 24%), *Centropyxis aculeata* (13%), *C. aerophila* (8%), *Euglypha rotunda* (8%), *Cryptodifflugia* sp. (7%) (figure 2). 36 species from the peat samples are used in the calibration model, so these species are included in the reconstruction of the water table depth dynamics. The other 28 testate amoebae species from peat samples do not participate in the calibration model. So their optima are not estimated and they are not included in the reconstruction. Although according to the literature [11-12] the most of these species are typical for minerotrophic mires.

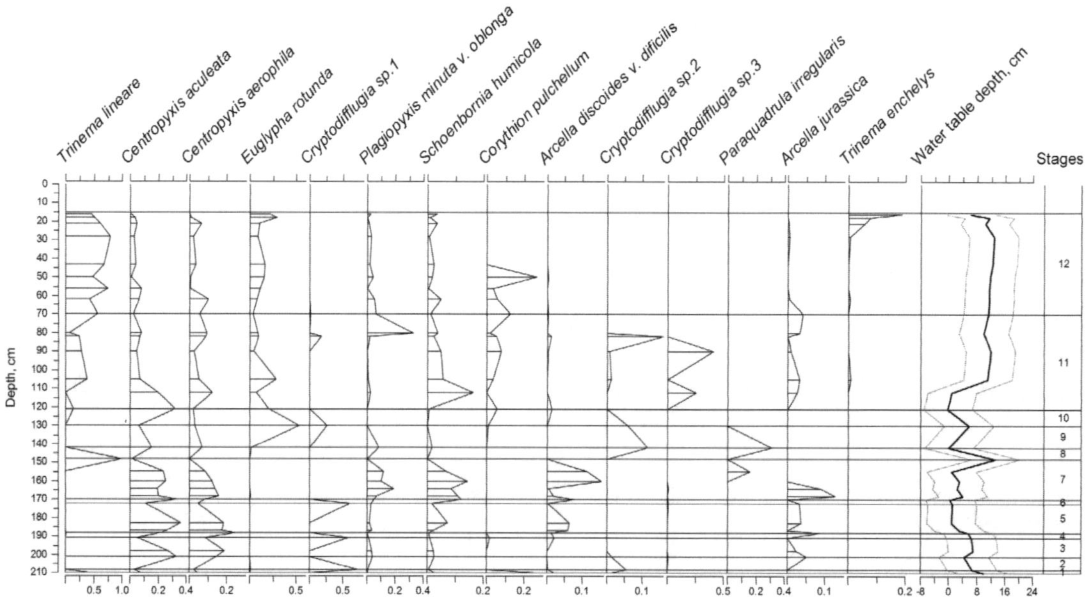


**Figure 2**. Testate amoeba relative abundance diagram showing all species with relative abundance of more than 10% in the peat core from the swamp Dolgon'koye.

The amoebae tests are well preserved in the studied peat core. The average number of species per sample is high (20 ± 5 SD). In the most part of the peat samples (21) more than 300 specimens of testate amoebae were counted, in 9 peat samples – more than 150 specimens and only in 2 peat samples (from the depth 170–169, 183–182 cm) less than 150 specimens were counted. It is important to note that Euglyphid species: *Trinema lineare, Euglypha rotunda, Corythion pulchellum*, – reached a high abundance throughout the depth of the peat deposits. Although according to research [13-14] the Euglyphid species are very rarely found in peat bogs and are characterized by low preservation of tests compared to other species. As our results showed, Euglyphid species seem to be well preserved in the minerotrophic peat. This fact further emphasizes the prospects of using rhizopod analysis in paleoecological studies of minerotrophic mires.

Twelve stages of the swamp development and its WTD dynamics are distinguished by the stratigraphically constrained clustering of testate amoebae assemblages in the peat core.

Stage 1. 210–208 cm (3105–3075 cal yr BP) – testate amoebae assemblages are dominated by *Trinema lineare* and *Corythion pulchellum* (dry condition indicators). They represent the low level of the swamp surface wetness (WTD is 10.0 cm).

Stage 2. 208–201 cm (3075–2980 cal yr BP) – this stage is characterized by increases in *Cryptodifflugia* sp.1. This species is accompanied by *Cyclopyxis eurystoma*, *C. eurystoma* v. *parvula*, *C. kahli*, *Plagiopyxis minuta* v. *oblonga*. A combination of these species indicates the low level of the swamp surface wetness (WTD is 6.8 cm).

Stage 3. 201–191 cm (2980–2835 cal yr BP) – species *Centropyxis aculeata* and *C. aerophila* dominate the testate amoebae assemblages. In addition, the assemblage contains dry indicators: *Plagiopyxis minuta* v. *oblonga* and *Cyclopyxis eurystoma.* It represents the period with slightly wetter conditions (WTD is 5.8 $\pm$ 1.8 cm).

Stage 4. 191–188 cm (2835–2780 cal yr BP) – *Cryptodifflugia* sp.1 dominates as well. With this species *Centropyxis aerophila, C. constricta* v. *minima, C. aculeata* are present in the assemblage. WTD is slightly increases up to 6.5 cm. Surface wetness seems to be a little less.

Stage 5. 188–172 cm (2780–2480 cal yr BP) – this stage is characterized by dominating of *Centropyxis aculeata* and *C. aerophila.* These species are accompanied with *C. cassis, Arcella discoides* v. *dificilis, A. jurassica, Schoenbornia humicola and Cyclopyxis eurystoma.* It indicates a tendency towards wetter conditions on the swamp surface (WTD is 3.1 $\pm$ 2.4 cm).

Stage 6. 172–170 cm (2480–2440 cal yr BP) – testate amoebae assemblages are dominated by *Cryptodifflugia* sp.1. There are also hydrophilic species: *Centropyxis aculeata, Arcella jurassica, Schoenbornia humicola*, and dry indicator: *Plagiopyxis minuta* v. *oblonga.* WTD is 1 cm.

Stage 7. 170–148 cm (2440–2280 cal yr BP) – species *Centropyxis aculeata* dominates. It represents wet conditions in the swamp surface (WTD is 2.2 $\pm$ 1.5 cm).

Stage 8. 148–142 cm (2280–2245 cal yr BP) – this stage is characterized by absolute dominance of the species *Trinema lineare*. It indicates dry period (WTD is 13.4 cm).

Stage 9. 142–130 cm (2245–2140 cal yr BP) – species *Paraquadrula irregularis* reached a high abundance. This species was not found in our study in modern conditions, but in accordance with another research [12] it is most typical for bryophyte tufts in calcareous fens. So it represents the period with the wettest conditions in the swamp surface (WTD is 0 cm).

Stage 10. 130–121 cm (2140–2030 cal yr BP) – this short stage is characterized by dominating of *Euglypha rotunda.* This species is accompanied with *Cryptodifflugia* sp.1, *Cryptodifflugia* sp.2, Centropyxis *aculeata, Trinema lineare* and *Corythion pulchellum.* Therefore, a combination of these species indicates the middle level of the swamp surface wetness (WTD is 5.9 cm).

Stage 11. 121–70 cm (2030–855 cal yr BP) – this stage is heterogeneous with different dominant species of testate amoebae. High percentages of species *Cenropyxis aculeata, Trinema lineare, Plagiopyxis minuta* v. *oblonga, Cryptodifflugia* sp.3 are recorded in this stage. Overall, a combination of dry and wet indicators represents the middle level of swamp surface wetness with a high degree of dispersion (WTD is 7.4 $\pm$ 5.5 cm).

Stage 12. 70–15 cm (855–215 cal yr BP) – the stage is characterized by high abundance of *Trinema lineare* and represents dry conditions in the swamp surface (WTD is 11.1 $\pm$ 1.9 cm). The last period of the stage (230–215 cal yr BP) is characterized by increases in *Trinema enchelys.* It indicates the tendency towards slightly wetter conditions on the swamp surface (WTD is 6.3 cm).

Our result of the reconstructed swamp surface wetness is in good agreement with the paleoclimatic data having been obtained by other researchers for the south of West Siberia region (figure 3). The wet conditions in the swamp coincide with the increase of climate humidity 2700 cal yr BP, revealed, on the one hand, in the research of Zakh with co-authors [7], who studied lake sediments of several lakes located about 1000 km in the north-west using multi-proxy approach (archeological, geological and pollen data). On the other hand, this event is supported by the results from the paleohydrological reconstruction using rhizopod analysis of the peat on the ombrotrophic bog located 550 km in the north-west [3]. Besides, in accordance with the other researches [3, 6-7] the climate between 2500 and 1900 cal yr BP was humid. Two of these researches [3, 7] were mentioned above, and the third [6] was carried out using pollen and diatom records from the sediments of the lake situated 400 km in the west. This corresponds to wet conditions in the studied swamp during this period of the Holocene. Than in one of the works [7] a short decrease in the climate humidity was revealed between 2300 and 2100 cal yr BP, and in the other work [3] dry conditions on the bog were detected between 2300 and 2200 cal yr BP and in 2000 cal yr BP. In our study, there are coincident brief decreases of the swamp surface wetness in 2280 and 2140 cal yr BP. The period of a drier climate between 1900 and 600 cal yr BP, revealed in other research [6], corresponds with dry conditions in the swamp surface at this time. This event also coincides with dry period revealed in the study [5] of the sediments from four lakes located about 400 km in the west, where paleoclimate was reconstructed based on the ostracod analysis. It also corresponds well to the some short dryings (1750, 1600, 1500, 1300–1200, 1100–1000 cal yr BP) on the bog surface in the study [3]. Although according to Zakh and co-author's data [7] the climate was humid, nevertheless in the period later than 1000–700 cal yr BP drier climatic conditions were shown as well. In the study [7], the increase of climate humidity was shown for the last 300 years, in the other [3] - between 550 and 350 years and in the third study [6] it was shown for the period 750-500 cal yr BP. Slightly wetter conditions in the swamp surface are observed between 230 and 215 cal yr BP in accordance with our research. On the whole, a high degree of consistency of the swamp surface wetness with the paleoclimatic data indicates a high sensitivity of the peat deposits in the studied swamp to climatic changes that occurred in the Holocene.

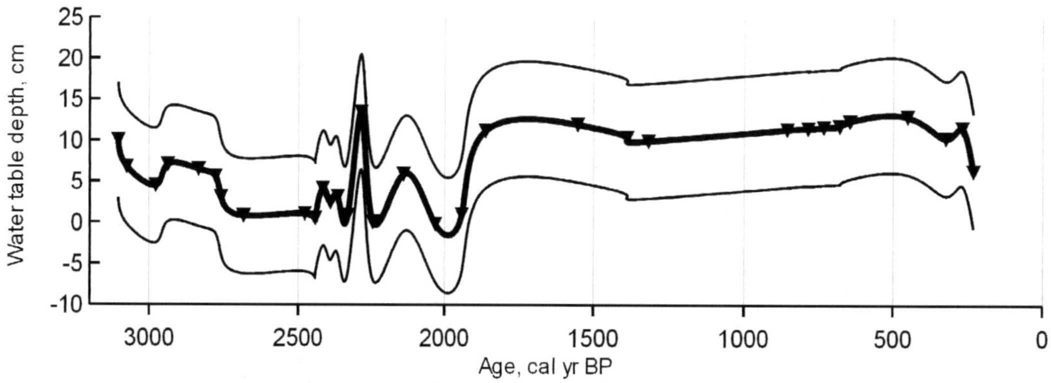


**Figure 3**. Testate amoeba based paleohydrology of the swamp Dolgon'koye.

## 5. Conclusion

Rhizopod analysis in the peat deposits of the swamp revealed well preserved testate amoebae assemblages with high numbers of tests and of species. Therefore, this provided a quantitative reconstruction of the water table depth during the swamp development. The observed changes in species composition of testate amoebae assemblages throughout the depth of peat deposits and oscillations of reconstructed water table depth are clearly the result of paleoenvironmental fluctuations in the process of the swamp development in the Holocene. Our study demonstrated applicability of rhizopod analysis in paleoecological studies of minerotrophic mires.

**Acknowledgements:**
This study was supported by the grant of the Russian Foundation for Basic Research (17-55-52020). We thank P.A. Blyakharchuk and S.V. Loyko for participating in the expedition and for collecting materials for the analysis.

## References

[1] Lamentowicz M, Slowinski M, Marcisz K, Zielinska M, Kaliszan K, Lapshina E, Gilbert D, Buttler A, Fialkiewicz-Koziel B, Jassey V E J, Laggoun-Defarge F and Kolaczek P 2015 *Quaternary Research* **84 (3)** pp 312–325
[2] Turner T E, Swindles G T and Roucoux K H 2014 *Quaternary Science Reviews* **84** pp 65–85
[3] Willis K S, Beilman D, Booth R K, Amesbury M, Holmquist J and MacDonald G 2015 *The Holocene* **25 (9)** pp 1425–1436
[4] Kremenetski K V, Velichko A A, Borisova O K, MacDonald G M, Smith L C, Frey K E and Orlova L A 2003 *Quaternary Science Reviews* **22** pp 703–723

[5] Khazin L B, Khazina I V, Krivonogov S K, Kuzmin Y V, Prokopenko A A, Yi S and Burr G S 2016 *Russian Geology and Geophysics* **57 (4)** pp 574–585
[6] Rudaya N, Nazarova L, Nourgaliev D, Palagushkina O, Papin D and Frolova L 2012 *Quaternary Science Reviews* **48** pp 32–42
[7] Zakh V A, Ryabogina N E and Chlachula J 2010 *Quaternary International* **220 (1-2)** pp 95–101
[8] Luzgin B N 1998 *Izvestiya of Altai State University* **(10)** pp 113–116
[9] Chambers F M, Blackford D W and Yu Z 2011 *Mires and Peat* **7** pp 1–10
[10] Hendon D and Charman D J 1997 *The Holocene* **7(2)** pp 199–205
[11] Chardez D 1965 *Bulletin de l'Institut Agronomique et des Stations de Recherche de Gembloux* **33** pp 307–341
[12] Opravilova V and Hajek M 2006 *Acta Protozoologica* **45** pp 191–204
[13] Swindles G T and Roe H M 2007 *Palaeogeography, Palaeoclimatology, Palaeoecology* **252** pp 486–496
[14] Mitchell E A D, Payne R J and Lamentowicz M 2008 *J. Paleolimnol.* **40** pp 603–618

IOP Conf. Series: Earth and Environmental Science **138** (2018) 012010 doi:10.1088/1755-1315/138/1/012010

# Polygonal patterned peatlands of the White Sea islands

**S A Kutenkov*[1], M N Kozhin[2,3,4], E O Golovina[5], E I Kopeina[4] and N V Stoikina[1]**

[1] Institute of Biology, Karelian Research Centre, Russian Academy of Sciences 11 Pushkinskaya St., 185910 Petrozavodsk, Karelia, Russia
[2] Lomonosov Moscow State University, 1 Leninskiye Gory, 119234, Moscow, Russia
[3] Kandalaksha State Nature Reserve 35 Lineynaya St., 184042, Kandalaksha, Murmansk Region, Russia
[4] The Polar-Alpine Botanical Garden-Institute of Avrorin, Kola Science Center, Russian Academy of Sciences, 184256, Kirovsk-6, Murmansk region, Russia
[5] Komarov Botanical Institute of the Russian Academy of Sciences, 2 Professor Popov St., Saint-Petersburg, 197376, Russia

*Email: effort@krc.karelia.ru

**Abstract.** The summits and slopes of some islands along the northeastern and northern coasts of the White Sea are covered with dried out peatlands. The thickness of the peat deposit is 30–80 cm and it is separated by troughs into gently sloping polygonal peat blocks up to 20 $m^2$ in size. On some northern islands the peat blocks have permafrost cores. The main components of the dried out peatlands vegetation are dwarf shrubs and lichens. The peat stratigraphy reveals two stages of peatland development. On the first stage, the islands were covered with wet cottongrass carpets, which repeated the convex relief shape. On the second stage, they were occupied by the xeromorphic vegetation. We suggest that these polygonal patterned peatlands are the remnants of blanket bogs, the formation of which assumes the conditions of a much more humid climate in the historical past. The time of their active development was calculated according to the White Sea level changes and radiocarbon dates from 1000–4000 BP.

## 1. Introduction

There exist islands covered with dried out peatlands in the White Sea, the appearance of which resembles the shape of ice-wedge polygonal mires of arctic lowlands. The peat deposit is separated by troughs into gently sloping polygonal peat blocks 4–20 $m^2$ in size. A layer of withered peat with a thickness of 30–80 cm repeats the convex relief shape covering the summits and slopes of the land. The islands are far beyond the zone of the Arctic polygon mires, and the ice wedges, which are typical for the polygon mires, appear to be always absent in the troughs of the peatlands. The xeromorphic vegetation, that covers the peatlands, consists mainly of dwarf shrubs and lichens. The purpose of this work is to give brief essay on the current vegetation, to survey the peat deposit stratigraphy and to reconstruct the Holocene history of these peculiar peatlands.

## 2. Study area

The White Sea is located in the northeastern part of Europe along the boundary of Fennoscandia. In the period from 2006 to 2016 more than 750 km of the coastline were observed between the mouths of the Ponoi and Kem' rivers (figure 1). All studied islands are located along the coast with a maximum distance of 14 km. The polygon patterned peatlands were found on 9 islands. Additionally peatland with hummocky-hollow surface pattern was observed on Veshnyak island (table 1).

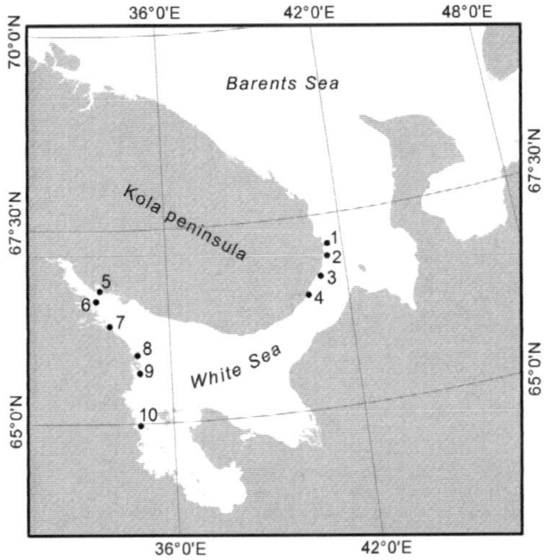


**Figure 1.** Study islands: 1 – Veshnyak, 2 – Unnamed island near the Goryainov island, 3 – Danilov, 4 – Sosnovetz, 5 – Stolbovaya Luda, 6 – Bol'shaya Srednyaya Luda, 7 – Sharapikha, 8 – Zelenaya Luda, 9 – Bol'shoi Rob'yak, 10 – Terroikha.

**Figure 2.** Polygon patterned peatland on Danilov island.

The surveyed territory is located within Belomorian fold belt – the southeastern part of the ancient Precambrian Fennoscandian shield [1]. Bedrocks dominated by acid gneisses and granite gneisses, lacking in almost all biogenic elements, some richer basic rocks cover smaller areas [2]. Islands in southern and western part are bare fractured rock outcrops whereas in the northeastern part, they have a thin layer of loose marine deposits [3].

**Table 1.** Study islands.

| Island | Geographic coordinates N | E | Altitude above sea level, m | Size, km² | Peatlands mineral basement |
|---|---|---|---|---|---|
| Veshnyak | 67°06'25" | 41°24'07" | 20.1 | 0.621 | soft sediments above rock |
| Unnamed near Goryainov | 67°01'40" | 41°20'48" | < 5.0 | 0.008 | soft sediments above rock |
| Danilov | 66°44'20" | 41°05'33" | 6.3 | 0.036 | soft sediments above rock |
| Sosnovetz | 66°29'22" | 40°40'59" | 16.2 | 0.491 | soft sediments above rock |
| Stolbovaya Luda | 66°40'42" | 33°47'02" | 19.1 | 0.024 | bare rock |
| Bol'shaya Srednyaya Luda | 66°36'03" | 33°41'04" | 15.5 | 0.050 | bare rock |
| Bol'shoi Rob'yak | 66°37'31" | 34°54'26" | 16.8 | 0.269 | bare rock |
| Sharapikha | 66°13'41" | 34°03'09" | 31.8 | 0.112 | bare rock |
| Zelenaya luda | 65°53'39" | 34°48'40" | 11.0 | 0.058 | bare rock |
| Terroikha | 65°02'40" | 34°59'25" | 22.5 | 0.232 | bare rock |

Among the most important factors in the history of the formation of coastal peatlands during the Holocene are the isostatic land uplift and oscillations of the White Sea level [4, 5]. The rate and sometimes the direction of vertical movements in different parts of the studied territory are different [6, 7]. Since the highest middle Holocene transgression (ca. 6250 BP, Tapes phase), the sea level along the west coast has dropped by 20–25 m, while in the northeast it has dropped to 10 m. The most distinct coast paleoline, corresponding to the sea level 4100 years ago traced at 17 m a.s.l., is on the western coast and it is 8 m a.s.l. on the northeastern coast. The modern uplift is 1–2 mm per year in the west and the fall is 1.4–2.5 mm in the northeast [7].

The climatic regime of the White Sea region may be described as intermediate between marine and continental types. The winters are relatively warm but lengthy whereas summers are short and cold. Additional specific climate features include significant cloudiness, high air humidity, abundant precipitation, and generally unstable weather conditions throughout the year. The intensive cyclonic activity and relatively rapid shifts of synoptic processes create high variability of the values of meteorological parameters [8]. The long-term mean annual air temperature along the studied area ranges from –0.67 in the northeast (Sosnovetz island weather station) to +1.4 °C (Kem' weather station) in the south-west. The annual precipitation is 343 mm (Sosnovetz) and 467 mm (Kem'), 71–75% of them fall in the warm period of the year [8].

The studied islands are mainly covered by heaths, which vary based on geographic location. The southern and western islands are situated within the north taiga subzone [9]. As a rule, *Empetrum hermaphroditum* is the most abundant species of these heaths, *Vaccinium vitis-idaea* and *V. uliginosum* are common. On the well-drained substrates *Arctostaphylos uva-ursi* and *Arctous alpina* are important. *Vaccinium myrtillus*, *Avenella flexousa,* and *Luzula pilosa* are confined to the snow-protected habitats. *Andromeda polifolia*, *Rubus chamaemorus,* and *Carex rarifolra* are significant in the paludified heathlands. Mosses (mainly *Pleurozium schreberi*), lichens (*Cladonia* spp., *Flavocetraria nivalis*, etc.) and liverworts (chiefly *Ptilidium ciliare*) are common. The peculiarity of these island heaths is the presence of some boreal, and even hemiboreal, plants (*Rubus saxatilis*, *Luzula pilosa*, *Dicranum polysetum*, etc.) and the absence of some arctic-montane or hypoarctic species (*Betula nana*, *Alectoria ochroleuca*, *Thamnolia vermicularis*, etc.), which are common on the northern islands. The other vegetation types of the southern and western islands are the supralittoral coastal meadows, salt marshes, fragments of mires, and pine forests, which occupy a negligible area.

The northern islands are lying within the south tundra (low arctic) subzone [9]. The Veshnyak island flat summit is mainly covered by chionophobous heath with lichens *Cladonia arbuscula* and *Flavocetraria nivalis* predominate. *Empetrum hermaphroditum*, *Betula nana* and prostrate *Salix glauca* prevail among the vascular plants. On the steep shore slopes there are *Empetrum hermaphroditum*, *Betula nana*, *Chamaepericlymenum suecicum* dominating heath and willow (*Salix glauca*, *S. lanata*) thickets. The Danilov and the small unnamed islands are entirely covered by dried out peat deposits with *Empetrum hermaphroditum* heath with abundance of *Rubus chamaemorus* (figure 2). The Sosnovetz island covered by peat deposits that have the form of a palsa peatland and in some places of a slightly differentiated plateau. The mounds and plateau are mainly covered by *Cladonia arbuscula*, *Flavocetraria nivalis* dominated heath with a high significance of *Rubus chamaemorus* and *Empetrum hermaphroditum*. *Eriophorum scheuchzeri*, *Carex aquatilis*, *C. rariflora* and *Sphagnum lindbergii*, *S. riparium* predominate in the hollows. The other vegetation types (mainly coastal meadows and herbs coenoses) of the northern islands are of little significance.

The polygon patterned peatlands are separate massifs on most of the islands. On some islands the polygonal structures are combined with extensive permafrost peat plateaus. On the Danilov island the polygonal relief turns into a more homogeneous plateau with permafrost at the northern foothill. The hilltop of Sosnovetz is entirely occupied by a permafrost peatland with an undulating relief, by the edges of which a polygonal structure is developing along the slopes and on the foothill. The studied peatlands on Veshnyak island are spotted by flat summits and they have hummocky-hollow relief.

## 3. Methods

### *3.1. Vegetation studies*

The peatlands vegetation was recorded with about 100 relevés, which were made in homogeneous stands from sample plots 25 m² in 2011–2016. The abundance of each taxon was recorded by means of a cover percentage scale. The location in mesorelief, nanorelief pattern, exposure and slope angle were recorded for each sample plot.

### *3.2. Peat studies*
Peat samples were taken manually from small pits to the full depth usually every 5 cm if a uniform layer or by visible limits if noticed. In Stolbovaya Luda and Bol'shaya Srednyaya Luda the thickness of samples was 2 cm. In Bol'shoy Rob'yak, Sharapiha, Zelenaya luda it was from 5 to 15 cm. For the most part of the islands the peat samples were taken only from one point, but on bigger islands they were taken from several sampling points. For the Danilov island we laid stratigraphical geological cross-section based on 5 sampling points through the whole peatland. Altogether 151 peat samples from 14 pits were analyzed in this study.

Botanical macrofossil analyses and degree of decomposition evaluation of samples were made with a microscope method at the precision of 5% in the Laboratory of mire ecosystem IB KarRC RAS in Petrozavodsk by N. Stoikina. The radiocarbon dating was performed in the conventional $^{14}C$ laboratories at the Geological Institute of the Russian Academy of Sciences (Moscow) (GIN) and Herzen State Pedagogical University of Russia (St.Petersburg) (SPb).

### *3.3 Meteorological data*
The main sources of meteorological data of Hydrometeorological Service in Russia are Raspisaniye Pogodi Ltd. (rp5.ru) and hydrometeorological database (http://ib.komisc.ru/climat) [10].

## 4. Results

### *4.1 Essay on the current vegetation*
The main components of the dried out peatlands vegetation are the dwarf shrubs (chiefly *Empetrum hermaphroditum*, also *Vaccinium uliginosum*, *V. vitis-idaea*, *Arctous alpina*), lichens (*Cladonia arbuscula*, *C. rangiferina*, *Flavocetraria nivalis*, *Cetraria islandica*, etc.) and *Rubus chamaemorus*. The moss floor is poorly developed usually and consists of *Dicranum elongatum*, less often *Polytrichum strictum*. Liverworts are common: *Ptilidium ciliare* grows mainly on the top of polygons; *Calypogeia* spp. and *Cephalozia* spp. are confined to the polygon slopes. There are deflation scars on the most windswept sites on the polygon surface, which are overgrown by *Ochrolechia frigida* and cup-shaped *Cladonia* species. As a rule, the troughs between polygons are occupied by the same dwarf shrubs and sometimes also by the same lichens. The narrowest troughs lack any vegetation. The mire hygrophilous species and the ice wedges, which are typical for the Arctic polygon mires [11], are always absent in the troughs of the dry polygon peatland. Another difference from palsa and polygon mires is the absence of *Andromeda polifolia*, *Carex rariflora*, *Eriophorum vaginatum*, *Sphagnum* spp., etc. in the communities of the top of the polygons. Such communities with some variations in structure and floristic compositions are typical for the dry peatlands of the studied islands, except for Veshnyak Island. On the latter island, the studied peatlands have hummocky-hollow relief. The hummocks 40–60 cm high are covered by heath with abundance of *Empetrum hermaphroditum*, *Rubus chamaemorus*, *Betula nana* and sometimes of lichens (*Cladonia* spp., *Cetraria islandica*); *Carex rariflora* is constant. The hollows are dry with mineral substrate or a thin peat layer on the bottom. There are scattered *Polytrichum* spp., lichens, liverworts and solitary *Salix glauca* or *S. lanata* in them.

### *4.2 Peat stratigraphy*
The peat deposits are dry. The fragments of permafrost are observed only on the northeastern islands (Danilov, Sosnovetz). It occurs in the central parts of peat blocks at the depth of 40 cm and penetrates deeper into the sand. The basal layer of the deposit is dense dark brown peat while the upper layer is more bulk, have lighter brown tint and permeated with living roots of shrubs.

The basal layer of the deposits is composed of cottongrass peat with decomposition degree 35% or more. Besides *Eriophorum*, remnants of Ericales, *Rubus chamaemorus*, *Salix*, *Betula*, *Spagnum* sect. *Cuspidata*, *Dicranum*, *Polytrichum* are common. The remnants of *Carex rariflora*, *C. aquatilis*, *Equisetum* and herbs are less seen. Almost pure cottongrass deposit was observed only once – at the foothill of Danilov island (figure 3a), where it was overlapped with a layer of logs, brought by the sea (not shown on the graph). In addition, the basal peat layer is frozen here; both of these factors prevent the penetration of shrub roots into the underlying deposit. In addition to the *Eriophorum*, there are remnants of *Salix*. The maximum thickness of the cottongrass basal peat is 25–30 cm (figure 3b, l). In some cases *Eriophorum* is not absolute dominant of basal peat, although it is present in significant amounts (figure 3c, h, k).

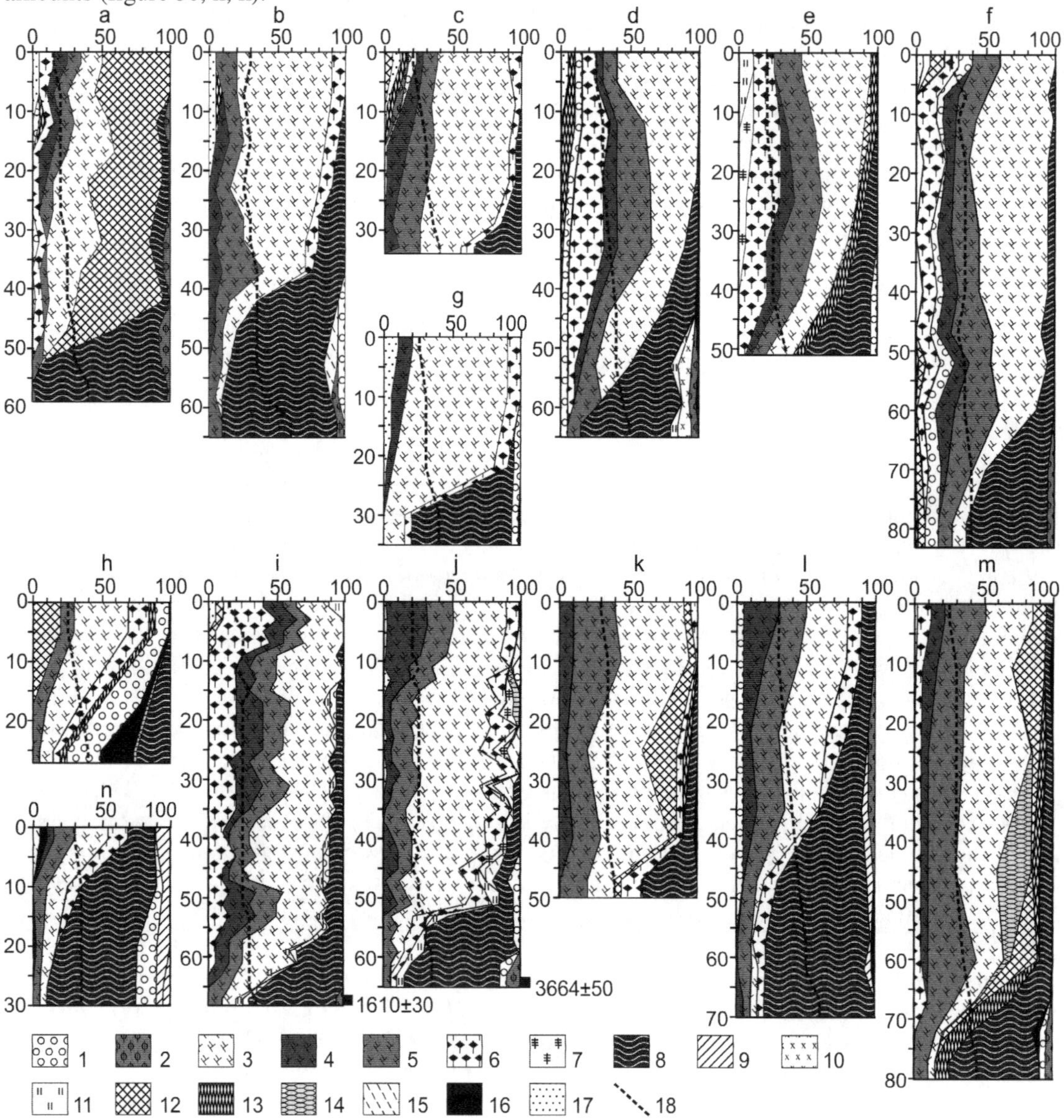


**Figure 3.** Peat macrofossil composition columns (a–n – Islands: a–c – Danilov (a – foothill, b – seaside slope, c – hilltop); d–f – Sosnovetz (d – shoreside slope, e – foothill, f – seaside slope); g – Unnamed island near Goryainov island; h, n – Veshnyak (hilltops); i – Stolbovaya Luda; j – Bol'shaya Srednyaya Luda; k – Bol'shoy Rob'yak; l – Sharapikha; m – Zelenaya luda. 1–17 – Macrofossils: 1 – *Betula nana*, 2 – *Salix*, 3 – *Empetrum*, 4 – *Vaccinium*, 5 – Ericales (undetected), 6 – *Rubus chamaemorus*, 7 – *Chamaepericlymenum suecicum*, 8 – *Eriophorum*, 9 – *Carex* (*C. rariflora* – i, l, n; *C. aquatilis* – i (bottom)), 10 – *Equisetum*, 11 – Herbs (undetected), 12 – *Dicranum*, 13 – *Polytrichum*, 14 – *Pleurozium*, 15 – Bryales (incl. *Loeskypnum badium* – b, i), 16 – *Sphagnum* (*S.* sect. *Cuspidata* – d, g, k, l; *S. lindbergii* – h; *S. riparium* – n; *S. capillifolium* – m), 17 – Lichens; 18 – Decomposition degree (%).

The cottongrass peat is overlapped by dwarf-shrubs one with decomposition degree of 20–35%. There are deposits with both sharp and slightly more smoothed boundary of dwarf-shrub peat. The most detailed study was carried out at Stolbovaya Luda and Bol'shaya Srednyaya Luda, the transition is relatively smooth on the first one, and on the second island *Eriophorum* content abruptly decreases from 65 to 10%. In the above mentioned Danilov island deposit with artificial boundary in the form of logs the transition of peat layers is quite sharp. *Empetrum* remnants is the dominant of dwarf-shrub peat layer, *Vaccinium* is also usual, some of the shrubs cannot be determined, but most likely they also belong to these genera. *Rubus chamaemorus* is common. Peculiar remnants are of *Dicranum* (probably, *D. elongatum*) and *Polytrichum* species, in some cases, the former plays an essential role (figure 3a). Small cottongrass content (5–10%) can be found close to the surface (up to 5–15 cm); in a few cases, it completely disappears at the depth of 25–30 cm (figure 3j, k). Sometimes there are remnants of *Chamaepericlymenum suecicum*. The thickness of the dwarf-shrub peat layer in the deeper deposits may reach up to 60 cm, in shallow – 15–30 cm.

Commonly, there is no cottongeass at all in the surface layer of peat as well as in the current communities. The other peat forming plants are supplemented with the remnants of lichens and hepatic mosses. The degree of decomposition decreases to 15%. The stratigraphic profile from the Danilov island demonstrates the stratification of these layers over the entire area of the peatland (figure 4).

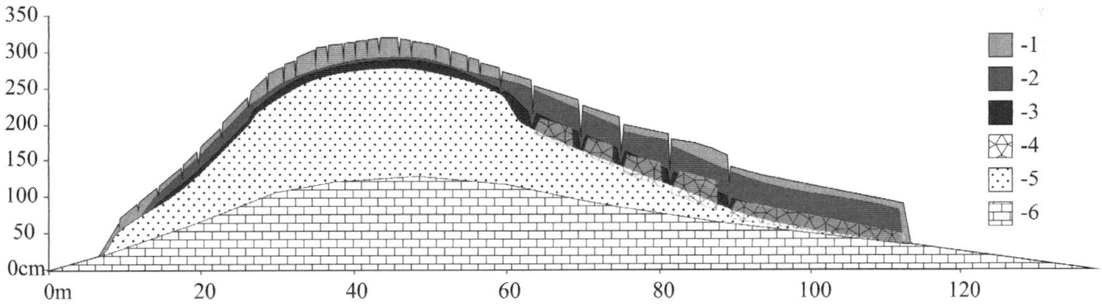


**Figure 4.** Stratigraphical geological cross-section (west-east direction) of the peatland on Danilov island. 1 – dwarf-shrub peat, 2 – dwarf-shrub peat with cottongrass, 3 – cottongrass peat, 4 – permafrost, 5 – sand, 6 – rock basement (sand/rock boundary is suppositive).

In addition to the typical polygonal patterned peatlands we studied some similar formations. For example, in figure 3 h, n, the stratigraphy of shallow peat deposits of the hummocky-hollow complex from Veshnyak island are given. The main features of the deposit structure are similar to those of the other deposits. The peculiarities are the presence of *Sphagnum lindbergii* remnants in the basal layer in one case and *Carex aquatilis* in the second one as well as the larger role of *Betula nana*.

There are two radiocarbon datings of basal cottongrass peat layers: from the islands Stolbovaya Luda 1610±30 BP (GIN-15057) and Bol'shaya Srednyaya Luda 3664±50 BP (SPb-1928).

## 5. Discussions

### *5.1 Paleocommunities*

The stratigraphy of the deposit of all polygonal patterned peatlands is rather similar. In all cases two main layers of peat are identified – the basal layer, with the degree of decomposition of 35–50%, consisting predominantly of the remnants of cottongrass and the overlapping layer of dwarf-shrub peat with a degree of decomposition of 20–30%, in which the remnants of cottongrass are found in small amounts close to the surface. Thus, the stratigraphy of the peat deposits identifies two stages of vegetation development. At the first stage the islands were covered by wet cottongrass carpets repeating the convex shape of the relief. In addition to *Eriophorum*, Ericales, *Salix*, *Betula nana*, *Sphagnum* sect. *Cuspidata*, *Carex rariflora*, herbs and some Bryales species were in paleocommunity

IOP Conf. Series: Earth and Environmental Science **138** (2018) 012010 doi:10.1088/1755-1315/138/1/012010

composition. In conditions of a slow water flow on flat plateaus, as on Veshnyak island, *Sphagnum* and sedges (including *Carex aquatilis*) were abundant.

Currently there are no such communities on the convex relief forms in the region. On some islands of the east of the Kola Peninsula on the mires with permafrost, some thawed hollows occupied by a carpet of cottongrass (*Eriophorum scheuchzeri*, *E. russeolum*, *E. media*) can still be found, but this relates only to the depressions of micro-relief. Mires of convex relief forms with mostly atmospheric nutrition are known as blanket bogs which global distribution today is confined to cool, extremely oceanic climate [12, 13, 14, 15]. The cottongrass is the main peat-forming plant of the blanket bogs in the oligotrophic (acid) conditions [12]. Based on our studies of the deposits, similar paleocommunities occupied large areas on some islands, often being the main type of vegetation in the historical past.

The second stage of peatlands vegetation development was its xeromorphisation. Cottongrass paleocommunities were changed by dwarf-shrub–moss and, later, in some places, by dwarf-shrub–lichen. Most species of cottongrass paleocommunities are absent in the modern flora of peatlands.

*5.2 Time frame*

The comparison of the current absolute hypsometric position of peatlands with the history of the White Sea level fluctuations during the Holocene allows us to determine to some extent the time of the beginning of the formation of peatland on the islands.

The sea level in the western part during highest middle Holocene transgression (ca. 6250 BP) exceeds present time level for 20–25 m [7] and all studied islands were below sea level or in the zone of active wave action. In the eastern part, the sea level was 10 m higher than the current and some of the islands were under water as well. The sea level experienced weak oscillation till 4000–4100 BP, when it reached 15–17 m [4, 7]. After that, the last significant regression of the sea started (Trivia-Ostrea stages transition). Finally, the White Sea level fell to an absolute 10 m only 2000 years ago. At that time, the lowest of the studied islands in the western part of the sea came to the surface. Thus for the most of the studied peatlands the age of origin cannot exceed 2000–4000 years. The time of peatland initiation on the Bol'shaya Srednyaya Luda (3664 ± 50 BP) is close to that assuming the sea level oscillations.

*5.3 Climatic envelopes*

The blanket bogs existence in the past in the White Sea region is a remarkable fact. Currently, the nearest active blanket bogs are located in the north of the central part of Norway [13, 14] which is more than 800 km to the northwest from the White Sea coast. The existence of blanket bogs is limited by narrow climatic envelopes [15, 16, 17]. There are several climatic indexes delineating blanket bog formation conditions. Further we compare them with modern conditions along the coast of the White Sea obtained by long-term observations at the weather stations.

The first one is the mean annual temperature (MAT) > −1°C [15]. In this study region the MAT is –0.67 °C (for Sosnovetz island) and 1.4 °C (for Kem' river estuary), thus, throughout the region, the MAT does not exceed the climatic limits for the blanket bogs. Accordingly [16] blanket bogs occur in Fennoscandia in areas where the annual temperature varies from 1.5 to 6.7 °C, but in their opinion, temperature variables here appear not to be as important climate determinants.

Another temperature parameter, the mean temperature of the warmest month (MTWA) varies from 8.5 (Sosnovetz) to 14.2 (Kem'), which corresponds to MTWA < 14.5 °C [15] and MTWA 10.0–14.1 °C [16]. The last important temperature parameter is continentality (yearly thermic interval, max.–min. of monthly mean temperatures); for Sosnovetz (21.0) and Kem' (27.2) it exceeds the upper climatic limit of 18.9 [16] due to low winter temperatures.

The second important group of parameters relates to precipitation. In accordance with [16] the mean annual precipitation (P) explained almost 60% of variation in blanket bog occurrence in Fennoscandia. The greatest discrepancy between the current climatic conditions of the White Sea area and the blanket bog regions is in much lower precipitation amount. Blanket bogs occur in Fennoscandia in the areas where P exceeds 1018 mm [16], whereas the long-term mean for Sosnovetz and Kem' is 2–3 time less.

The moisture index, MI (ratio of P to equilibrium evapotranspiration) might be expected to capture the distribution of peatlands better than merely P because the latter neglects the large differences in evaporative demand between climates at different latitudes and with differing degrees of cloudiness [17]. MI for blanket bogs occurrence area is > 2.1. The gross evaporation for Ponoi river watershed is 236 mm and for Kem' river is 356 mm, the calculation model demonstrates similar values [18]. So, accordingly to gross evaporation MI is 1.45 for Ponoi–Sosnovetz and 1.19 for Kem' area, which is also significantly lower than MI of blanket bogs region. It turns out that with the same evaporation rate, to create conditions for the initiation of blanket bogs, the amount of precipitation should increase by 152 mm (30%) in the Ponoi–Sosnovetz and 322 mm (43%) in the Kem' area.

### *5.4 Past climatic possibilities for blanket bog initiation*

For the eastern Fennoscandia there are a number of spore-pollen diagrams for organic deposits of mires, on the basis of which the quantitative characteristics of the paleoclimate were determined using the information statistical method based on a nonlinear statistical relationship of current climate conditions with the surface (subrecent) pollen spectra [19]. One of such pollen-based climatic reconstructions was made for Lovozero mire from inner part of Kola Peninsula [20]. It allows tracing possible climatic circumstances for the initiation of blanket bog formation during the Holocene.

The most favorable conditions for the blanket bogs inception initiations were during the climatical optimum in the Atlantic period (higher amount of precipitation, lower rate of climate continentality). At that time, the most massive formation of blanket bogs over Great Britain started [12, 15]. However, the studied islands were hidden by the waters of the White Sea.

At the transition of the Atlantic and Subboreal periods 4500 BP there was a significant global cooling during which palsa mires started to form on the Kola Peninsula about 4000 BP [20]. The next warming maximum occurred in 3500 BP. Precipitation exceeded the current level by 50 mm, the temperature - by 2 °C, the climate continentality decreased. The radiocarbon date of the basal cottongrass peat layer from Bol'shaya Srednyaya Luda (3664 ± 50 BP) exactly corresponds to this climate shift. We suppose that during this warming in the middle of the Subboreal the blanket bogs started to form on the White Sea islands. At the same time, the extension of blanket bogs on the British islands is traced [15], as well as peatland formation on the fells of Finnish Lapland [21]. So, the climate of Northern Europe at this time contributed best to the expansion of the blanket bogs.

However, some of the studied islands at this time were still below the sea level. The last warming event occurred in the lesser climatic optimum of the Medieval period, with the maximum about 1000 years ago. Radiocarbon dating from the Stolbovaya Luda (1610±30 BP) is of a slightly older age. The amount of precipitation exceeded then the current level by 75 mm, temperature - by 1–2 °C [20]. By that time all the surveyed islands already appeared. The above comparative climate data show that an increase in MAT and MTWA by 1–2 degrees would not exclude these areas beyond the climatic optimum for the development of blanket bogs. The continentality decline is due to a greater increase in winter temperatures; it could even approach the upper optimum limit.

To the greatest extent, the climatic optimum is not matched by precipitation. The above calculations show that at the same rate of evaporation, the amount of precipitation required for blanket bog initiations should have exceeded the current ones by 150–300 mm. The calculated precipitation rate [20] refers to the inner part of the Kola Peninsula; perhaps the increase in precipitation at the seashore was significantly greater. Moreover, the islands might have inherent lower evaporation due to higher permanent air humidity on the sea coast in comparison with the inner regions of the continent. The average annual air humidity at Sosnovetz is 89%, whereas at the Kanevka weather station located in the interior of the mainland at 80 km from Sosnovetz is 82%. In the warm months the difference is even more significant (89% and 78%). There are 60–70 days with fog during year in the northern part of the White Sea and in the summer months fogs occur every 3 days [22]. These factors reduced the evaporation and led to an increase of the substrate humidity.

It is known that the climate of the White Sea is unstable both on a short-term and on a long-term scale [8, 23]. Currently, the increase in precipitation is observed, i.g. for Kem' annual precipitation grows from 425 (1968 yr) to 514 (2014 yr), i.e. at 89 mm, significant growth is also observed at the other stations along the White Sea [23]. It is caused by the intensification of cyclonic activity and the increase in the intensity of precipitation [24]. Thus, it can be assumed that the amount of precipitation on the White Sea coast in the past could also be significantly higher than the current level and conditions for the blanket bogs development might well exist.

The warm period was replaced by a "small Ice Age" with an extreme at about 800 BP, when precipitation and temperature were lower than the present-day values by 50 mm and by 3–4 °C (2 °C in July) and the climate continentality increased. That cooling event led to degradation of Sphagnum on palsas and the formation of dwarf shrub and lichen communities, which still persist in the area [20]. We assume that at the same time the blanket bogs of the White Sea islands might dry up.

Indirectly, it is indicated by the position of the lower boundary of the peatlands at about 10 m a.s.l. in the western part of the sea. The slopes of the islands appeared less than 1000 years ago were never occupied by cottongrass communities. In the northeastern islands, which are currently experiencing subsidence [7], the peatlands extend to supralittoral grass communities.

The existence of permafrost fragments in blocks of some northern peatlands is an interesting fact. There is a suggestion that the permafrost as waterproof horizon, can contribute to the blanket bogs development [25]. However, the studies of climatic envelops [15, 16] show that palsa mires and blanket bogs in the north of Europe have a clear thermal disjunction, which means that they cannot be at one time in one place. Suggesting, the development of permafrost in the peatlands of the White Sea islands occurred later or during the changes in their plant communities. As in the case of Fennoscandia palsa mires [26], the existence of permafrost is currently maintained in them due to strong winds, which are drying the peat surface and keeping it clear of snow in winter.

## 6. Conclusion

The drying out polygonal patterned peatlands of summits and slopes of the White Sea islands are the evidence of blanket bog existence during a more humid climate periods of 1000–4000 years ago. The amount of precipitation on the White Sea coast required for their existence at that time had to exceed significantly the previously assumed value. To the present time, these peatlands stopped their development and the erosion processes are prevailing. This study provides only a general idea about the history of their development. Relict peatlands of the islands of the White Sea are quite unique formations and undoubtedly require further investigations.

## Acknowledgments

We are grateful to O L Kuznetsov and A I Maximov, who gave advice on the field work. E G Ershova and O I Smushlyaeva for participation in the field work. N A Kozhin, P N Vaganov for technical and organizational assistance in the field work. G A Tyusov for his help in the map designing. Andrew Schutt, improved the English. This research was funded from grants of RFBR № 16-05-00644 and state assignment of IB KarRC RAS № 0221-2017-0048. The authors are thankful to anonymous reviewer who provided valuable comments and suggestions.

## References

[1] Slabunov A I, Stepanova A V, and Sibelev O S 2008 Geological characteristics *Rupestrian Landscapes of the White Sea Karelian Coast: Natural Characteristics, Economic Utilization, Conservation* ed A N Gromtsev (Petrozavodsk: Karelian Research centre of RAS) chapter 1.2 pp 8–15

[2] Systra Y I 2003 Geologic characteristics *Biotic Diversity of Karelia: Conditions of Formation, Communities and Species* ed A N Gromtsev et al (Petrozavodsk: Karelian Research centre of RAS) chapter 1.2 pp 7–12

[3] Pozhilenko V I, Gavrilenko B V, Zhirov D V and Zhabin S V 2002 *Geology of mineral areas of the Murmansk Region* (Apatity: Kola Science Centre RAS) p 359

[4] Elina G A 2008 Mires and paludified lands characteristics *Rupestrian Landscapes of the White Sea Karelian Coast: Natural Characteristics, Economic Utilization, Conservation* ed A N Gromtsev (Petrozavodsk: Karelian Research centre of RAS) chapter 2.1 pp 28–50
[5] Zaretskaya N E, Shevchenko N V, Simakova A N and Sulerzhitsky L D 2011 *Geochronometria* **38 (2)** 116–127
[6] Kaplin P A and Selivanov A O 1999 *Changes in the Sea Level of Russia and the Coast Development: Past, Present, Future* (Moscow: GEOS) p 288
[7] Koshechkin B I 1979 *The Holocene Tectonic of the Eastern Part of Baltic Shield* (Leningrad: Nauka) p 158
[8] Filatov N N, Pozdnyakov D V, Johannessen O M, Pettersson L H and Bobylev L P 2005 *White Sea. Its Marine Environment and Ecosystem Dinamics Influenced by Global Change* (Chichester: Springer-Praxis Publ.) p 472
[9] *Vegetation of the European Part of the USSR* 1980 ed E M Lavrenko et al (Leningrad: Nauka) p 236
[10] Novakovskiy A B and Elsakov V V 2014 *Data Science Journal* **13** 57–63
[11] Boch M S, Gerasimenko T V and Tolchelnikov Yu S 1971 *Botanical Journal* **56 10** 1421–1435
[12] Moore P D, Merryfield D L and Price M D R 1984 The Vegetation and Development of Blanket Mires *European Mires* ed P D Moore (London: Acad. Press) 203–235
[13] Moen A 1999 Vegetation *National Atlas of Norway* (Hønefoss: Norwegian Mapping Authority) 200
[14] Pajunen H 2005 Mires *The Physical Geography of Fennoscandia* ed M Seppälä (Oxford: Oxford University Press) 77–95
[15] Gallego-Sala A V, Charman D J, Harrison S P, Li G and Prentice I C 2016 *Clim. Past* **12** 129–136
[16] Parviainen M and Luoto M 2007 Climate envelopes of mire complex types in Fennoscandia *Geogr. Ann.* **89 (2)** 137–151
[17] Gallego-Sala A V, Clark J, House J I, Orr H G, Prentice I C, Smith P, Farewell T and Chapman S J 2010 *Clim. Res.* **45** 151–162
[18] Salo Yu A, Nazarova L E and Balagansky A F 2016 *Trans. Karelian Research Centre RAS* **9** 95–101
[19] Klimanov V A 1976 *Vestn. Moscow State University* Ser 5 Geography **2** 92–98
[20] Elina G A, Arslanov Kh A, Klimanov V A and Usova L I 1995 *Bot. J.* 80 (3) 1–16
[21] Luoto M and Seppälä M 2000 *The Holocene* **10 (2)** 229–241
[22] Vasilyev L Yu and Vodovozova T E 2010 Climate *The White Sea system **1** Natural environment of the catchment area of the White Sea* ed I A Nemirovskaya and V P Shevchenko (Moscow: Scientific World) pp 16–39
[23] Tolstikov A V 2016 *Variability of the Surface Layer Temperature of the White Sea* (Moscow: GEOS) p 212
[24] Filatov N N, Nazarova L E, Georgiev A P, Semenov A V, Anciferova A R, Ozhigina V N and Bogdan M I 2012 *Arctic: Ecology and Economy* **2(6)** 80–93
[25] Neshatayeva V Yu and Neshatayev V Yu 2006 *Proc. Int. Symp. Mire Ecosystems in Nothern Europe: Diversity, Dynamics, Carbon Balance, Resources and Conservation (Petrozavodsk: aug. 30 – sept. 2 2005)* (Petrozavodsk: Karelian Research centre of RAS) pp 181–192
[26] Seppälä M 2011 *Quaternary Research* **75 (2)** 366–370

IOP Conf. Series: Earth and Environmental Science **138** (2018) 012011 doi:10.1088/1755-1315/138/1/012011

# Diversity of mire massif types in the boreal zone of European Russia

**O L Kuznetsov**

Laboratory of Mire Ecosystems, Institute of Biology, Karelian Research Centre, Russian Academy of Sciences, Petrozavodsk, Russia

Email: kuznetsov@krc.karelia.ru

**Abstract.** In Russia, mire massif type is the principal structural unit for descriptions of the diversity of regional mire ecosystems of various ranks, vegetation mapping, and decision-making on the use of mires. The classification of mire massifs is based on various criteria and indicators. The botanical-geographical classification of mire massifs of the boreal zone of European Russia is four-tiered, and includes 22 types gathered in groups, subgroups and three classes. For most of the types their characteristic associations and diagnostic species are stated.

## 1. Introduction

The three major structural levels distinguished for mire ecosystems are mire sites (habitats), mire massifs and mire systems. The central object for the study and classification of mire ecosystems in Russian mire science is mire massifs, which develop in isolated depressions or on watersheds. Studies on mire massif classification have a more than a century-long history. There are several classification approaches, the main ones being the trophic, botanical-geographical, geomorphological and hydrological [1-3]. In each of these classifications the units of different ranks are determined not only by the principal traits on which they are based, but also by additional indicators.

The most thoroughly elaborated and widely used in many spheres of science and practice are botanical-geographical classifications of mire massifs, especially in various kinds of vegetation mapping [3-6]. The sets and distribution patterns of plant communities and their combinations (composite habitats) in massifs are one of the basic criteria for such classifications. Their geographic affiliation is also taken into account and reflected in the names of types. The botanical-geographical classification of mire massifs developed for European Russia by T. Yurkovskaya [3] is made up of 28 types, grouped into 5 classes with groups and subgroups distinguished within them. The boreal (taiga) zone of this region has only 20 types belonging to 4 classes, including the forest mires class. Mire massif types in this classification are rather broad, and in the vegetation map of Europe they are even more general [2]. The above classifications were designed to meet the purposes of small-scale mapping (1: 1000 000 and smaller) of mire or vegetation types for very extensive regions. This always means that quite a broad set of mire types are generalized within one composite type, even though they vary widely in the composition and structure of the plant cover and in ecological characteristics. Massifs smaller than 100 (or even 1000) hectares are thus overlooked. When describing (assessing) the diversity of the plant cover and mire ecosystems of smaller areas, such as individual protected areas, there is no need to classify all the mire massif types identified here into known generalized (broad) regional types. The more important task is to state their detailed typology, features of the plant cover structure, and nature conservation value. Identification of new, narrower massif types is necessary when assessing the specifics of mire ecosystem diversity at the regional level, as demonstrated in a number of publications [7, 8].

## 2. Material and Methods

This endeavor was based on materials from long-term field surveys of mire ecosystems performed by the Laboratory of Mire Ecosystems, Institute of Biology KarRC RAS in Karelia and some adjacent regions in the European north of Russia [9, 10]. The typology of mire massifs was developed relying on the basic classification principles set out in a number of papers [3-6]. It is for the first time in a mire classification that the syntaxa (associations) the most characteristic of the massif's central part, as well as some diagnostic species discriminating them from similar types are provided for a majority of massif types.

## 3. Results and Discussion

A more detailed botanical-geographical classification of mire massif types has been developed for the north of European Russia within its boreal zone (table 1). It is four-tiered, and includes 22 types grouped into three classes, with groups and subgroups of types distinguished within them. Classification by T. Yurkovskaya [3] was taken as the backbone, with additions and amendments made at the level of subgroups and types of mire massifs. Also, variants were distinguished within some types. Names of massif types are taken over from Yurkovskaya's classification if their scope is also retained. Digital codes are introduced for all classification levels, permitting quick reference to the massif's status and easy modification and expansion of the classification.

New or modified types and their variants are named after the dominant communities in the central parts of the massifs or after their trophic status, as well as regional affiliations. For the Sphagnum bogs class (1), for instance, subgroups of massifs with compound and homogenous plant cover structure were distinguished, and a subgroup of bogs with *Sphagnum angustifolium* and *S. magellanicum* was singled out (table 1). New types of massifs were identified within several groups: for dystrophic suboceanic Sphagnum bogs (1.1.1) these are the Lapland (1.1.1.2) and North-East European (1.1.1.3) types. A finer division was suggested for Sphagnum poor fens (1.4.1), which are differentiated into 4 types, two of them with variants. There is also a new typology for aapa mires (class 2). The mesotrophic and eutrophic variants, which are distinct in their flora and the spectrum of plant communities, were distinguished within the Karelian (2.1.1.2) and Onega-Pechora (2.1.2.2) types. The class of herb and herb-brown moss mires (boreal subgroup 3.1.1) now comprises 4 types (Yurkovskaya's classification had only one). Mire massifs with a sparse tree stand (canopy closure 0.2-0.3) and abundant moss cover, which is the main producer component of the communities, were included in the classification.

Forest mires (wetland forests and swamps) with a canopy closure of 0.4 or more are not considered here. They constitute a separate class [3], and their typology is based on different indicators [11]. There are occasional palsa mires in the utter north of the forest zone, but their typology needs to be further developed together with other types of mires of the tundra zone.

The syntaxa (associations) that are the most typical for the massif's central part and define their overall appearance are provided for a majority of types and their variants (table 2). Associations were determined by the ecological topology method relying on dominant or diagnostic species for the main layers of the community [12, 13] Also, diagnostic species enabling clear discrimination of some massif types were identified (table 2). They are mostly species found at the margins of their distribution range due to climatic or biogeographic factors, such as some subatlantic (*Calluna vulgaris, Trichophorum cespitosum, Carex livida, Sphagnum cuspidatum, S. rubellum*), hypoarctic (*Carex rotundata, Sphagnum lindbergii*), East European and Siberian (*Chamaedaphne calyculata, Carex omskiana, Betula humilis*) species.

IOP Conf. Series: Earth and Environmental Science **138** (2018) 012011 doi:10.1088/1755-1315/138/1/012011

**Table 1.** Classification of mire massif types of taiga zone the European Part of Russia.

| **CLASS** and group of mire massifs (after: [3], with additions) | Subgroup of mire massifs (after: [3], with additions) | Mire massif types and their variants (after: [3]* and in own interpretation) |
|---|---|---|
| 1 | 2 | 3 |
| **1. BOREAL SPHAGNUM BOGS**<br>1.1 European suboceanic Sphagnum raised bogs (distrophic) | 1.1.1 Liverwort-lichen-Sphagnum ridge-hollow raised bogs with secondary pools | 1.1.1.1. Calluna-lichen-Sphagnum- liverwort ridge-hollow raised bogs of the White Sea coastal region*<br>1.1.1.2 Lapland Empetrum-lichen-Sphagnum-liverwort ridge-hollow raised bogs<br>1.1.1.2a inland<br>1.1.1.2b Barents Sea coastal<br>1.1.1.3. North-East European Empetrum-lichen-Sphagnum-liverwort ridge-hollow raised bogs |
| 1.2 North-West European Sphagnum raised bogs | 1.2.1 Sphagnum ridge-hollow raised bogs with *Calluna vulgaris* and *Sphagnum fuscum* | 1.2.1.1 Karelian dwarf shrub-cloudberry-Sphagnum ridge-hollow raised bogs<br>1.2.1.1a North-Karelian*<br>1.2.1.1b Middle-Karelian*<br>1.2.1.2 West-Russian dwarf shrub-Sphagnum ridge-hollow raised bogs* |
| | 1.2.2 Homogenous Sphagnum bogs with *Sphagnum angustifoli-um, S. fuscum, S. magellanicum* | 1.2.2.1 North-West European pine-cottongrass-Sphagnum bogs*<br>I.2.2.2 North-West European cottongrass-Sphagnum bogs |
| I.3 North-East European Sphagnum raised bogs | 1.3.1 Sphagnum ridge-hollow raised bogs with *Chamaedaphne calycylata* and *Sphagnum fuscum* | 1.3.1.1. Chamaedaphne-cloudberry-Sphagnum ridge-hollow raised bogs of the Onega-Pechora Region* |
| | 1.3.2 Homogenous Sphagnum bogs with *Sphagnum angustifoli-um, S. fuscum, S. magellanicum* | 1.3.2.1. North-East European pine-cottongrass-Sphagnum bogs*<br>1.3.2.2. North-East European cottongrass-Sphagnum bogs |
| 1.4 North-East European oligotrophic Sphagnum mires (poor fens) | I.4.1 Homogenous grass-Sphagnum poor fens | 1.4.1.1 Low sedge-dwarf shrub-Sphagnum (*Sphagnum papillosum, S. balticum, S. majus*) poor fens<br>1.4.1.1a Trichophorum-dwarf shrub-Sphagnum poor fens<br>1.4.1.1b Scheuchzeria-cottongrass- Sphagnum poor fens<br>1.4.1.2 Tall sedge-Sphagnum (*S. fallax, S. angustifolium, S. lindbergii*) poor fens |

| | | |
|---|---|---|
| | I.4.2 Grass-Sphagnum compound poor fens | 1.4.2.1 Low sedge-dwarf shrub-Sphagnum (*Sphagnum papillosum, S. balticum, S. majus*)-liverwort poor fens with hummock-flark microrelief<br>1.4.2.2 Dwarf shrub-grass-Sphagnum (*Sphagnum fuscum, S. jensenii, S. majus, S. lindbergii*) string-flark poor fens |
| | 1.4.3. Wood-herb-Sphagnum poor fens | 1.4.3.1. Pine-dwarf shrub-sedge-Sphagnum poor fens<br>1.4.3.2. Birch-pine-herb-Sphagnum poor fens |
| **2. GRASS-SPHAGNUM-BROWN MOSS FENS (AAPA MIRES)**<br>2.1. North European aapa mires | 2.1.1.North-West European aapa mires (Fennoscandian aapa mires) | 2.1.1.1. Lapland aapa mires with dwarf shrub-cloudberry-Sphagnum strings and herbs flarks*<br>2.1.1.2. Karelian aapa mires with *Betula nana-Molinia caerulea*-Sphagnum strings and herbs and herb-brown moss flarks*<br>2.1.1.2a mesotrophic<br>2.1.1.2b eutrophic |
| | 2.1.2. North-East European aapa mires | 2.1.2.1. Forest tundra aapa mires with dwarf shrub-cloudberry-Sphagnum strings and herb-moss flarks*<br>2.1.2.2. Aapa mires of the Onega-Pechora Region with dwarf shrub-Trichophorum-Sphagnum strings and herbs and herb-moss flarks*<br>2.1.2.2a mesotrophic<br>2.1.2.2b eutrophic |
| **3. HERB AND HERB-BROWN MOSS FENS**<br>3.1.East European herb and herb-brown moss fens | 3.1.1. Boreal herb and herb-brown moss fens | 3.1.1.1. Herb and shrub-herb floodplain and lacustrine shore mesoeutrophic (ME) and eutrophic (E) fens<br>3.1.1.1a Sedge fens<br>3.1.1.1b Herb and shrub-herb fens<br>3.1.1.2. Sedge-brown moss fens without spring water recharge<br>3.1.1.3. Herb-moss fens with spring water recharge |

This classification permits a more accurate determination of the affiliation of many syntaxa, as well as harvestable, rare and red-listed species to certain massif types. Thus, the most productive populations of *Menyanthes trifoliata, Comarum palustre* occur in floodplain and lacustrine shore herb and shrub-herb massifs (3.1.1.1). High cranberry yields are likely to be found in grass-moss poor fens (1.4.1) and cottongrass-Sphagnum raised bogs (1.2.2.2 & 1.3.2.2). A majority of vascular plant and moss species red-listed in Russia and boreal regions of its European part inhabit in spring-fed eutrophic (3.1.1.3) and aapa mires (2.1).

This classification will be further augmented, details on the distribution of individual types in the region will be added, and comparisons will be drawn to classifications of massif types for North European countries in order to produce a common classification for the entire Northern Europe.

**Table 2**. Characteristic syntaxa and diagnostic species of some types of massifs and their variants.

| Mire types | Syntaxa | Species |
|---|---|---|
| 1.1.1.1 | *Calluna vulgaris – Cladonia spp.* (ridges), *Scheuchzeria palustris –Sphagnum lindbergii, Trichophorum cespitosum – Hepaticae* (hollows) | *Calluna vulgaris, Carex rariflora, Sphagnum tenellum, S. capillifolium* |
| 1.1.1.2 | *Empetrum hermaphroditum – Cladonia* spp. (ridges), *Eriophorum vaginatum - Sphagnum lindbergii, E. vaginatum-Hepaticae* (hollows) | *Eriophorum russeolum, Carex rotundata* |
| 1.1.1.3 | *Empetrum hermaphroditum – Cladonia* spp. (ridges), *Eriophorum vaginatum - Sphagnum lindbergii, E. vaginatum-Hepaticae* (hollows) | absence of *Calluna vulgaris, Rhynchospora alba, Sphagnum tenellum* |
| 1.2.1.1 | *Chamaedaphne calyculata – Sphagnum fuscum* (ridges), *Scheuchzeria palustris –Sphagnum majus , Eriophorum vaginatum –S. balticum* (hollows) | *Calluna vulgaris, Trichophorum cespitosum, Sphagnum tenellum* |
| 1.2.1.2 | *Chamaedaphne calyculata – Sphagnum fuscum* (ridges), *Eriophorum vaginatum –S. balticum, Scheuchzeria palustris –Sphagnum cuspidatum* (hollows) | *Sphagnum cuspidatum, S. rubellum* |
| 1.2.2.1 | *Pinus sylvestris – Ledum palustre – Sphagnum angustifolium* | *Calluna vulgaris* |
| 1.2.2.2 | *Chamaedaphne calyculata – Eriophorum vaginatum - S. angustifolium* | *Carex pauciflora, Sphagnum papillosum* |
| 1.3.1.1 | *Chamaedaphne calyculata – Sphagnum fuscum* (ridges), *Scheuchzeria palustris –Sphagnum majus* (hollows) | absence of *Calluna vulgaris, Rhynchospora alba, Sphagnum tenellum* |
| 1.3.2.1 | *Pinus sylvestris – Ledum palustre – Sphagnum angustifolium* | absence of *Calluna vulgaris* |
| 1.3.2.2 | *Chamaedaphne calyculata – Eriophorum vaginatum - S. angustifolium* | absence of *Calluna vulgaris* |
| 1.4.1.1 | *Andromeda polifolia - Trichophorum cespitosum+ Carex pauciflora -– Sphagnum papillosum+S. balticum* | |
| 1.4.1.1a | *Andromeda polifolia - Trichophorum cespitosum – Sphagnum papillosum+S. balticum* | *Sphagnum compactum, S. pulchrum* |
| 1.4.1.1b | *Scheuchzeria palustris+Carex rostrata – Sphagnum majus* | *Sphagnum jensenii* |
| 1.4.1.2 | *Carex lasiocarpa – Sphagnum fallax, C. rostrara - S. angustifolium* | *Sphagnum subsecundum* |
| 1.4.2.1 | *Andromeda polifolia - Trichophorum cespitosum+ Carex pauciflora – Sphagnum papillosum+S. balticum* (carpets), *Carex limosa - Sphagnum majus +Hepaticae* (hollows) | *Drosera anglica, Rhynchospora alba* |
| 1.4.2.2 | *Chamaedaphne calyculata – Sphagnum fuscum* (ridges), *Carex limosa +Menyanthes trifoliata – Sphagnum jensenii* (flarks) | |
| 1.4.3.1 | *Pinus sylvestris – Chamaedaphne calyculata – Carex lasiocarpa -Sphagnum angustifolium* | |
| 1.4.3.2 | *Betula pubescens+ Pinus sylvestris – Carex lasiocarpa+Menyanthes trifoliata - Sphagnum angustifolium* | *Salix myrtilloides, Calliergon cordifolium* |

| | | |
|---|---|---|
| 2.1.1.1 | *Empetrum hermaphroditum+Rubus chamaemorus – Sphagnum fuscum* (strings), *Trichophorum cespitosum – Campylium stellatum* (carpets), *Carex limosa+Menyanthes trifoliata* (flarks) | *Carex rariflora, C. rotundata Eriophorum russeolum Loeskypnum badium* |
| 2.1.1.2 | *Carex lasiocarpa –Sphagnum fuscum* (hummocks), *Molinia caerulea- S. papillosum, Carex lasiocarpa –Sphagnum warnstorfii* (strings), *Carex limosa –Menyanthes trifoliata, Carex livida – Scorpidium scorpioides* (flarks) | |
| 2.1.1.2a | *Carex lasiocarpa –Sphagnum papillosum, Molinia caerulea- S. papillosum* (strings), *Carex limosa – Menyanthes trifoliata* (flarks) | *Calluna vulgaris, Trichophorum cespitosum, Nymphaea candida, Selaginella selaginoides* |
| 2.1.1.2b | *Carex lasiocarpa –Sphagnum warnstorfii, Molinia caerulea- S. warnstorfii,* (strings), *Carex livida – Scorpidium scorpioides, Carex limosa – S. scorpioides* (flarks) | *Juniperus communis Carex dioica, C. livida, Potentilla erecta, Sphagnum subfulvum, Scorpidium trifarium* |
| 2.1.2.1 | *Empetrum hermaphroditum+Rubus chamaemorus – Sphagnum fuscum* (strings), *Carex rariflora – Sphagnum lindbergii, Carex rotundata – Warnstorfia exannulata* (flarks) | absence of *Calluna vulgaris, Carex lasiocarpa, Scheuchzeria palustris* |
| 2.1.2.2 | *Carex lasiocarpa –Sphagnum magellanicum, Trichophorum cespitosum –S. papillosum, Trichophorum alpinum –S. warnstorfii* (strings), *Carex limosa – Menyanthes trifoliata, Carex limosa –Sphagnum jensenii* (flarks) | absence of *Calluna vulgaris, Molinia caerulea, Sphagnum subfulvum* |
| 2.1.2.2a | *Carex lasiocarpa –Sphagnum magellanicum, Trichophorum cespitosum –S. papillosum* (strings), *Carex limosa – Menyanthes trifoliata, Carex limosa –Sphagnum jensenii* (flarks) | *Eriophorum russeolum* |
| 2.1.2.2b | *Trichophorum alpinum –S. warnstorfii, Betula nana - Carex lasiocarpa –S. warnstorfii* (strings), *Carex limosa – Scorpidium scorpioides* (flarks) | *Betula humilis, Carex omskiana* |
| 3.1.1.1a | *Carex diandra – Comarum palustre, C. lasiocarpa - Comarum palustre, C. aquatilis, C. cespitosa, C. omskiana* | *Carex chordorrhiza, C. buxbaumii, C. panicea, Cicuta virosa, Lathyrus palustris, Calliergon giganteum* |
| 3.1.1.1b | *Salix lapponum – Carex rostrata, Equisetum fluviatile - Comarum palustre, Phragmites australis* | *Salix pentandra, S. cinerea, S. phylicifolia, Thyselium palustre, Hamatocaulis vernicosus* |
| 3.1.1.2 | *Carex lasiocarpa – Scorpidium scorpioides, C. limosa –S. scorpioides, C. lasiocarpa - Warnstorfia exannulata* | *Carex chordorrhiza, C. panicea, C. livida, Menyanthes trifoliata, Juncus stygius, Hammarbya paludosa, Sphagnum contortum, S. teres, S. obtusum* |

IOP Conf. Series: Earth and Environmental Science **138** (2018) 012011 doi:10.1088/1755-1315/138/1/012011

| | | |
|---|---|---|
| 3.1.1.3 | *Carex lasiocarpa –Bistorta major –Sphagnum warnstorfii, Molinia caerulea –Campylium stellatum, Trichophorum alpinum – Scorpidium cossonii* | *Salix myrsinites, S. rosmarinifolia, Saxifraga hirculus, Stellaria crassifolia, Epipactis palustris, Paludella squarrosa, Bryum pseudotriquetrum* |
| 3.1.1.4 | *Betula pubescens+Picea abies – Pharagmites australis – Sphagnum warnstorfii, Alnus glutinosa –Carex elongata, Pinus sylvestris –Equisetum palustre+Bistorta major – Sphagnum warnstorfii* | *Carex appropinquata, Listera ovata, Cypripedium calceolus, Helodium blandowii* |

**Acknowledgments**
The work was carried out under state order to IB KarRC RAS № 0221-2014-0035.

**References**

[1] Moen A 1995 *Gunneria* **70** 11–22
[2] Rybnicek K and Yurkovskaya T 1995 *Gunneria* **70** 67–72
[3] Yurkovskaya T K 1992 *Geography and Cartography of Mire Vegetation of the European Russia and Neighbouring Territories* (Saint-Petersburg: Komarov Botanical Institute RAS) p 256
[4] Tsinserling Yu D 1938 *Vegetation of mires* In: Vegetation in the USSR vol 1 (Moscow-Leningrad: Academy of Sciences of the USSR) pp 355–428
[5] Kats N Ya 1948 *Mire Types of the USSR and Western Europe and Their Geographical Occurence* (Moscow: Geographgis)
[6] Kats N Ya 1971 *Mires of the World* (Moscow: Nedra)
[7] Kuznetsov O, Tokarev P, Kutenkov S, Antipin V and Lindholm T 2014 *Rep. of the Finn. Environ. Inst.* **40** 131–45
[8] Kuznetsov O 2017 *Diversity of rich and intermediate fen types of taiga zone Russian European North* In: Proc. of the "VIII Meetings in Memoriam of Ekaterina Alexeevna Galkina" (Saint-Petersburg: Publishing house "LETI") pp 56–9
[9] Elina G A and Kuznetsov O L 2006 *The history and current status of mire research in Karelia: 55 years to Mire Ecosystems Laboratory* In: Proc. of Int. Symp. "Mire Ecosystems in Northern Europe: Diversity, Dynamics, Carbon Balance, Resources and Conservation" (Petrozavodsk: Karelian Research Centre RAS) pp 11–34
[10] Kuznetsov O 2015 *The main results of Karelian mire science school investigations* In: Int. symp. "Mires of Northern Europe: Biodiversity, Dynamics, Management" (Petrozavodsk: Karelian Research Centre RAS) pp 47–8
[11] Kutenkov S A and Kuznetsov O L 2013 *Diversity and dynamics of forested mires and paludified forests on the European North of Russia* In: Diversity and Dynamics of Forest Ecosystems in Russia vol 2 (Moscow: Publishing house KMK) pp 152–204
[12] Kuznetsov O L 2005 *Proc. of Karelian Res. Centre of Russ. Acad. Sci.* **8** 15–46
[13] Kuznetsov O L 2012 *The Finn. Environ.* **38** 133–42

# Soils in seasonally flooded forests as methane sources: A case study of West Siberian South taiga

**S Yu Mochenov[1], A I Churkina[1,2], S F Sabrekov[2,3], M V Glagolev*[1,2,3,4], D V Il'yasov[2,4], I E Terentieva and S S Maksyutov[5]**

[1] Faculty of Soil Science, Moscow State University, 1 Leninskie gory, Moscow 119991, Russia
[2] Water Problems Institute of RAS, 3 Gubkina, Moscow 119991, Russia
[3] UNESCO Department "Environmental Dynamics and Global Climate Changes", Yugra State University, Khanty-Mansiysk, 628012, Russia
[4] Institute of Forest Science Russian Academy of Sciences, Moscow Region, Russia
[5] Center for Global Environmental Research, National Institute for Environmental Studies, Nishi Odori Street, Tsukuba 305-8506, Japan

*Email: m_glagolev@mail.ru

**Abstract.** In this study, we measured the methane and carbon dioxide fluxes by static chamber method from the soil of periodically flooded forests under different water table levels (WTL) in West Siberian south taiga (Tomsk oblast, Russia) in summer seasons of 2016 and 2017 years. The study shows that seasonally flooded forests may become a methane source when the WTL increases up to 15-45 cm below the surface. The fluxes of methane from soil were from -0.08±0.07 to 9.3±0.8 mg·m-2·h-1, from 0.05±0.04 to 0.14±0.13 mg·m-2·h-1, from -0.03±0.02 to 5.4±0.2 mg·m-2·h-1 depending on various WTL in different seasonally flooded forests in 2017.

## 1. Introduction

Methane is an important greenhouse gas, which contributes to climate warming. The largest natural methane sources in the boreal latitudes are wetlands [1]. By now many researches have been carried out in those areas [2, 3, 4, 5], but still there are issues concerning other probable natural methane sources. The soils of periodically flooded forests may be one of them. The assumptions that forest soils affect the atmospheric methane balance, not only through the soil sink but also due to episodic high emissions in wet conditions were repeatedly given in publications [6, 7, 8]. The area and the period of temporal forest flooding may be probably estimated by hydrological model that uses meteorological information of a certain year. However, what part of periodically flooded forests area even under optimal conditions may become a methane source? Apparently, this question can not be solved by one-time measurements of the methane fluxes. Indeed the forest may be not under optimal (for emission) hydrothermal conditions at the time when a researcher carries out his measurements and instead of emission, it may reveal methane consumption. The aims of our study are to measure the methane flux at the soil surface and the profile of methane concentration in soil. It is obvious that at the same surface the methane flux concentration profile is to have a qualitative difference when: (i) there are only methanotrophs in the soil and (ii) when there are methanogens, under the methanotrophs layer, that may provide sufficient methane emission in the optimal conditions.

In this paper, we studied seasonally flooded forests as potentially significant methane sources in Western Siberia. Seasonally flooded forests can be methane sources or sinks depending on the season and precipitations [8]. Megonigal and Guenther [9] point out that upland soils which are mainly considered as methane sinks may also produce methane in microsites during brief periods of the soil moisture shift.

## 2. Materials and methods

### *2.1 Research sites*

The $CH_4$ and $CO_2$ flux measurements from the soil of seasonally flooded forests were carried out in Bakcharsky district of Tomsk oblast in the south taiga zone of Western Siberia during July and August 2016-2017 (figure 1).

**Figure 1.** Test sites.

The measurements were made at five test sites:

1) Seasonally flooded forest nearby Bakcharskoe bog (sites WF/RB, Wet Forest, Wet Forest_2);
2) "Ryam" or forested bog (pine-shrub-sphagnum ecosystem) at the Bakcharskoe bog border (site Low ryam_Bakcharskoe bog_1, 2);
3) Seasonally flooded forest within the drained bog (site Wet Forest_3);
4) Seasonally flooded forest within the floodplain of Bakchar river tributary (site Wet Forest_Polynyanka);
5) Ridge-hollow-lake complex of the Bakcharskoe bog (site Bakcharskoe bog, Plotnikovo).

The detailed site description is shown in the table 1. To understand the interannual variability, methane and carbon dioxide fluxes from soils have being measured for 3 years (2015-2017) at the sites WF/RB, Wet Forest, Wet Forest_2 The study of the WTL effect on methane and carbon dioxide fluxes from soils were carried out at the sites Wet Forest_3 and Wet Forest_Polynyanka in 2017. Additionally, the methane and carbon dioxide fluxes from soils were measured at the sites Low ryam_Bakcharskoe bog_1, 2 and the measurements at Bakcharskoe bog and Plotnikovo sites were made to compare the results with methane fluxes from typical mires.

**Table 1.** Description of test sites.

| Point | Coordinates | | Characteristics | | |
|---|---|---|---|---|---|
| | Latitude | Longitude | pH | EC | The most abundant woody species [a)] |
| T.Plo.H.1.3 | 56.81345 | 82.85368 | 4.43-5.60 | 30-115 | C.c., An.p., S.s. |
| fp_2 | 56.96189 | 82.51528 | 6.89-7.280 | 680-901 | B.p., C.v. |
| fp_3 | 56.96181 | 82.51528 | 7.01-7.16 | 523-623 | B.p., C.v. |
| fp_5 | 56.96128 | 82.51528 | 6.58-7.91 | 83-394 | B.p., C.v. |
| fp_6 | 56.96128 | 82.51533 | 6.69-7.31 | 508-809 | B.p., C.v. |
| fp_7 | 56.96131 | 82.51536 | 7.00-7.62 | 508-809 | B.p., C.v. |
| fp_8 | 56.96125 | 82.51542 | 6.91-7.20 | 855-880 | B.p., F.r., A.p. |
| fp_9 | 56.96128 | 82.51544 | n.d.[b)] | n.d.[b)] | B.p., A.p., C.a. |
| Tr. PWF | 56.83127 | 82.85285 | 5.48-6.26 | 92-191 | B.p. |
| Tr. PWF_2 | 56.83128 | 82.85150 | n.d.[b)] | n.d.[b)] | B.p. |
| WF_BD_01 | 56.87308 | 82.86350 | 5.03-5.17 | 44-78 | P.s., B.p. |
| WF_BD_02 | 56.87308 | 82.86356 | 4.54-4.78 | 70-83 | P.s., B.p. |
| WF_BD_1 | 56.87317 | 82.86386 | 5.31-6.26 | 101-390 | P.s. |
| WF_BD_2 | 56.87297 | 82.86381 | 5.28-7.56 | 150-289 | B.p., P.s. |
| WF_BD_3 | 56.87300 | 82.86383 | 6.28-6.28 | 130-130 | B.p., P.s. |
| WF_BD_4 | 56.87319 | 82.86381 | 5.10-5.09 | 152-160 | P.s. |
| WF_BD_5 | 56.87314 | 82.86381 | 4.74-5.50 | 64-131 | P.s. |
| WF_BD_6 | 56.87317 | 82.86383 | 4.46-4.64 | 165-166 | P.s. |
| WF_BD_7 | 56.87319 | 82.86367 | 4.27-4.66 | 130-143 | P.s. |
| WF_BD_8 | 56.87322 | 82.86358 | 4.29-5.48 | 86-157 | P.s. |
| Tr.WF/RB_1 | 56.83169 | 82.85122 | 4.31-4.61 | 79-90 | B.p., P.s. |
| Tr.WF/RB_2.1 | 56.83149 | 82.85133 | 4.63-4.63 | 193-193 | B.p., P.s. |
| Tr.WF/RB_2.2 | 56.83150 | 82.85133 | 5.76-5.76 | 296-296 | B.p., P.s. |
| Ryam_1 | 56.85467 | 82.85156 | 3.99-4.13 | 62-64 | P.s. |
| R_1 | 56.87217 | 82.84978 | 3.10-4.22 | 76-101 | P.s. |
| R_2 | 56.87206 | 82.84981 | 3.83-3.83 | 106-106 | P.s. |
| R_3 | 56.87206 | 82.84983 | 3.80-3.94 | 186-205 | P.s. |

[a] C.c. – *Chamaedaphne calyculata*, An.p. – *Andromeda polifolia*, S.s. – *Sphagnum* sp., B.p. – *Betula pendula*, C.v. – *Carex vesicaria*, F.r. – *Festuca rubra*, A.p. – *Aegopodium podagraria*, C.a. – *Calamagrostis arundinacea*, P.s. – *Pinus sylvestris*

[b] WTL was deeper than 60 cm.

*2.2 Methods*

The $CH_4$ and $CO_2$ soil flux measurements were made by a static chamber method [10] as it was described in [11]. The exposure time was 45–60 min depending on the water table level (WTL) for the methane. The column temperature in chromatograph was 35°C. The volume of injected methane sample was 2-3 ml. Every gas sample from a syringe was analyzed in 3-4 replicates; the average concentration was used for the flux calculation.

The carbon dioxide concentrations were analyzed during 1.5-5 hours after the gas sampling using an infrared gas analyzer DX6100-01 (RMT Ltd., Russia). It was previously calibrated using standard $CO_2$ gases with 0, 365, 390 ppm concentrations of ±0.02–0.04 ppm accuracy (National Institute for Environmental Studies, Japan). Before each measurement, the base line of analyzer was monitored using a mixture of air with $CO_2$ (357 ± 5 ppm; "MONITORING", Russia).

Moreover, the following environmental parameters were measured at each site: water table level, soil temperatures at depths of 0, 5, 15, 45 cm below the surface and 2 m above it using temperature loggers TERMOCHRON iButton DS 1921G ("DALLAS Semiconductor", USA), pH and electrical conductivity using Hanna HI-98129 Combo ("Hanna Instruments", USA).

The methane and carbon dioxide flux rates were calculated by linear regression with the weights and by non-linear (exponential) regression for emission and consumption, respectively, using chamber headspace concentrations versus measuring time [12]. To analyze the data, we used Wilcoxon rank sum test in MATLAB R2009a, regression analysis, Spearman rank correlation coefficient (P=0.05).

## 3. Results

The results of the individual $CH_4$ and $CO_2$ flux measurements from soils in 2015 are described in the article Glagolev et al. published in this Proceedings Book, in 2016 year they were published in [13], the results of the measurements in 2017 year are shown in the appendix. Based on these data, the weighted average value of $CH_4$ and $CO_2$ fluxes at different WTL were calculated for each site (table 2). Weighted average value and weighted error are used in the text below unless the otherwise is mentioned.

**Table 2.** The averaged $CH_4$ and $CO_2$ flux values at different sites

| $CH_4$ flux, $mg \cdot m^{-2} \cdot h^{-1}$ | | $CO_2$ flux, $mgC \cdot m^{-2} \cdot h^{-1}$ | | WTL, cm |
|---|---|---|---|---|
| Mean | Error | Mean | Error | |
| Bakcharskoe bog, Plotnikovo | | | | |
| 30.7 | 2.5 | 416 | 62 | 0 |
| Wet Forest_Polynyanka | | | | |
| 20.0 | 0.9 | 2 | 17 | -15 |
| 9.3 | 0.8 | 98 | 83 | 0 |
| 0.63 | 0.04 | 1 | 5 | 2.5 |
| 0.06 | 0.10 | 465 | 44 | 5 |
| -0.028 | 0.015 | 439 | 253 | 15 |
| -0.016 | 0.035 | 333 | 93 | 45 |
| -0.084 | 0.067 | - | - | 75 |
| WF/RB. Wet Forest. Wet Forest_2 [a)] | | | | |
| 0.05 | 0.04 | 204 | 26 | n.d.[b)] |
| 0.06 | 0.07 | 9 | 64 | 30 |
| 0.11 | 0.03 | 211 | 65 | 35 |
| 0.08 | 0.09 | 3 | 25 | 47 |
| 0.14 | 0.13 | 416 | 154 | 60 |
| Wet Forest_3 | | | | |
| 5.5 | 0.2 | 2 | 26 | 0 |
| 0.45 | 0.35 | 159 | 42 | 2 |
| 0.25 | 0.04 | 71 | 13 | 7 |
| 2.2 | 0.8 | 67 | 8 | 10 |
| 0.23 | 0.53 | 9 | 45 | 15 |
| 4.8 | 0.4 | 345 | 54 | 20 |
| 0.04 | 0.03 | 295 | 22 | 30 |
| -0.03 | 0.02 | 146 | 160 | 45 |
| Ryam. Bakcharskoe bog_m . Ryam. Bakcharskoe bog_h [a)] | | | | |
| 0.002 | 0.015 | 2 | 26 | 43 |
| 0.06 | 0.03 | 160 | 16 | 65 |
| -0.01 | 0.04 | 2 | 2 | 102 |

[a)] sites were joined because they were close to each other and had the same water source;
[b)] WTL was deeper than 60 cm.

Monitoring measurements of methane and carbon dioxide fluxes were carried out within the wet forest nearby the Bakcharskoe bog during 2015-2017 summer seasons (sites WF/RB, Wet Forest, Wet Forest_2; figure 2). The methane emission was found at the Wet Forest site in 2017, while the methane consumption was observed at this site in 2015-2016.

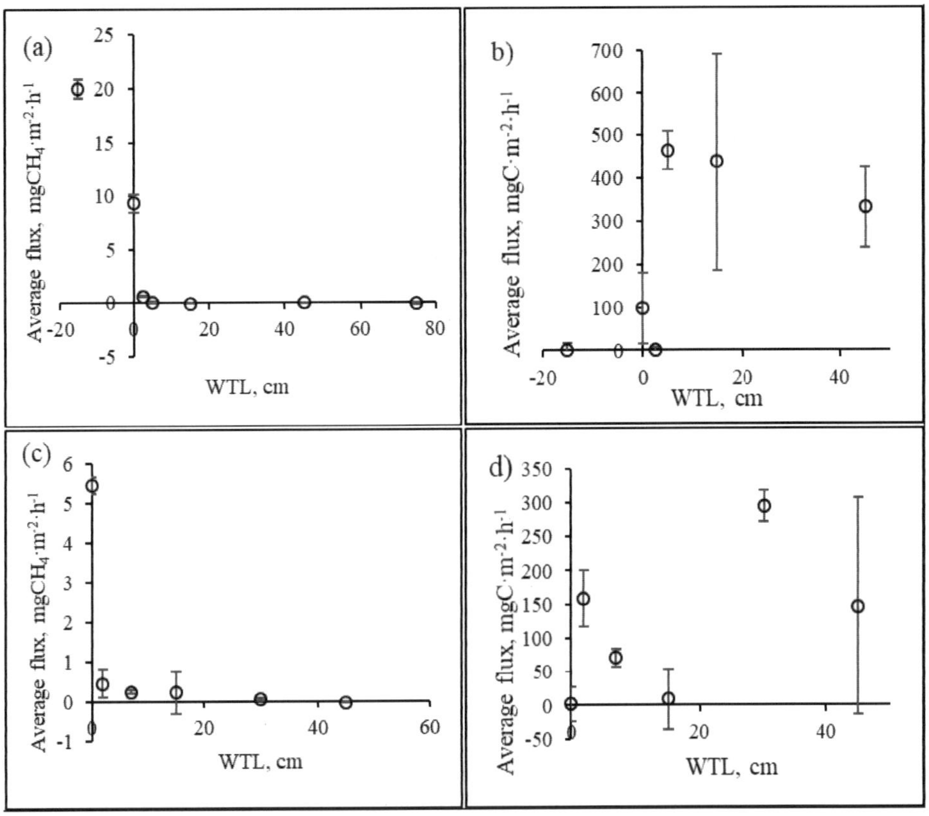


**Figure 2.** Averaged $CH_4$ and $CO_2$ fluxes under the different WTL: a) $CH_4$, Wet Forest_Polynyanka; b) $CO_2$, Wet Forest_Polynyanka; c) $CH_4$, Wet Forest_3; d) $CO_2$, Wet Forest_3. Only fluxes with n>1 were included.

To compare WTL controls on the $CH_4$ and $CO_2$ fluxes, emission measurements under the different WTL were carried out at the Wet Forest_Polynyanka and Wet forest_3 sites. The maximum methane flux was found at the seasonally flooded forest close to the Bakchar river tributary (Wet Forest_Polynyanka site) when the water level was 15 cm above the ground surface. The $CH_4$ fluxes sharply decreased when the WTL lowered beneath the ground (Spearman rank correlation coefficient r=-0.96). The average methane consumption by soils was observed when the WTL reached 15 cm below the ground level (figure 2.a).

At the Wet Forest_3 site $CH_4$ fluxes were lower comparing to the Wet Forest_Polynyanka site. However, there was a strong relationship with the WTL as well in the first case (Spearman rank correlation coefficient r=-1). The average methane consumption by soils was observed when the WTL reached 45 cm below the ground, which was three times lower than 15 cm at the Wet Forest_Polynyanka site (figure 2.c).

The WTL had another impact on $CO_2$ flux rates. The carbon dioxide flux reached its maximum at the WTL below 5 cm in the seasonally flooded forest within the floodplain of the Bakchar river tributary (Wet Forest_Polynyanka; figure 2.b). Spearman rank correlation coefficient (r=0.75) did not indicate a strong positive correlation of the carbon dioxide flux and the WTL.

The influence of the WTL increase on the average $CO_2$ flux was not obvious at the Wet Forest_3 site (figure 2.d). There was no positive correlation between the carbon dioxide flux and the WTL since the Spearman rank correlation coefficient (r=0.49) was lower than the critical value (r=0.85). The $CH_4$ and $CO_2$ fluxes from hollows of the ridge-hollow-lake complex of The Bakcharskoe bog were 30.7±2.5 and 416±62, respectively.

The methane concentration profiles in soils were made based on the measurements in 2016 and 2017 years (figure 3). At the relatively high WTL (about 10 cm below the surface; figure 3.a for 2016), the methane concentration increased immediately just below the ground level. When the WTL was deeper than 10 cm, the $CH_4$ concentration increased from the top soil layer to the water level (figure 3.a for 2017 year; figure 3.b). The $CH_4$ concentration decreased in this direction at the only one site (figure 3.c) where the $CH_4$ consumption was observed.

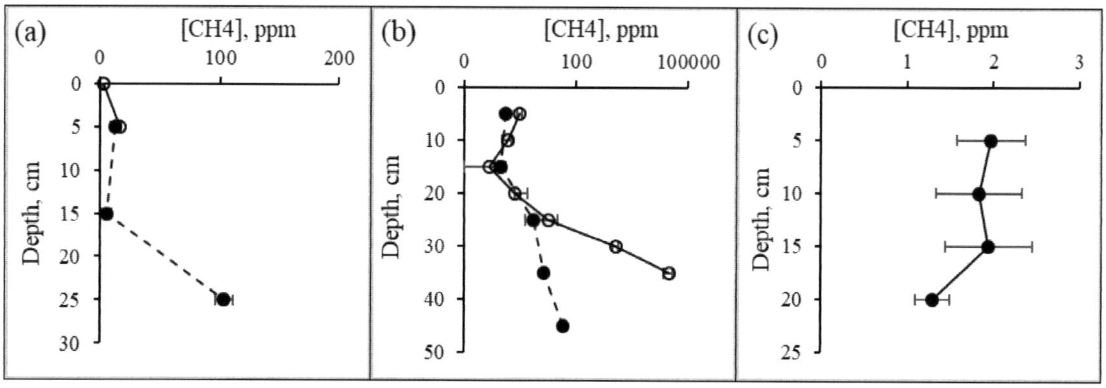


**Figure 3**. $CH_4$ concentration profiles in soil at the sites Tr.WF/RB_1 (a), Tr.PWF (b), fp_2 (c). The data from 2016 is marked by the solid line and the data from 2017 is marked by dashed lines.

## 4. Discussion

The typical methane concentration profiles in soils are shown in the figure 3. Generally, three types of concentration profiles were observed. The first one is at the relatively high WTL (about 10 cm below the soil surface; figure 3.a for 2016), at which the concentration increases immediately below the soil surface. The diffusive flux from the soil was roughly estimated using the data from 2016 season as an example. The $CH_4$ concentration gradient was 93÷116 mgC/m$^4$ (figure 3.a for 2016). The diffusion coefficient in the atmosphere was $D_O = 0.077$ m$^2$/h at 19 °C [14]. The ratio of the diffusion coefficient in the soil ($D_S$) to $D_O$ varied from 0 to about 0.3 depending on the air-filled porosity [15]. Thus, we took $D_S/D_O \approx 0.15$ for further calculations (this value was used just for rough estimation because the influence of the air-filled porosity on the diffusion coefficient is nonlinear). Finally, we estimated the diffusive flux from the soil as $(93 \div 116) \cdot 0.077 \cdot 0.15 \approx 1.1 \div 1.3$ mgC·m$^{-2}$·h$^{-1}$. Field chamber measurements conducted at the Tr.WF/RB_1 site in 2016 indicated flux median of 3.1±1.4 mgC·m$^{-2}$·h$^{-1}$. Taking into account the measurement error and calculation inaccuracy, the correspondence between calculated and measured fluxes may be estimated as satisfying. Actually, the diffusive flux may be lower than the total flux as the gas is emitted not only by diffusion but also by convection (for example, under pressure increase during methanogenesis).

When the WTL was deeper (about 25 cm below the surface), two other types of concentration profiles were found. The profile with gradual decrease of methane concentration with depth was observed in figure 3.c. We supposed that this profile was formed when there was no methanogenic activity or it was low in the underlying soil layers. The methane flux was roughly estimated as it was shown above for the profile in the figure 3.a for 2016 year. On the contrary with the first situation, the flux was negative, i.e. methane consumption was observed. It varied from -0.02 mgC·m$^{-2}$·h$^{-1}$ to 0 (we accepted the actual soil moisture as it was on 26 Jul 2016). The chamber measurements indicated $CH_4$ fluxes varying from -0.06±0.04 to 0.01±0.06 mgC·m$^{-2}$·h$^{-1}$ at the fp_2 site during this day in 2016.

The third type of profiles was characterized by the methane concentration decrease at the top layer with high methanotrophic activity and by its further increase related to higher methanogenic activity below the WTL (figure 3.b for 2016). The methane concentration decrease in the surface layer indicated the methane consumption from the atmosphere that was confirmed by chamber measurements. The theoretical calculations mentioned above estimated flux rates varying from -0.39

to -0.07 $mgC \cdot m^{-2} \cdot h^{-1}$ at the Tr.PWF site, while the direct chamber measurements showed fluxes varying from -0.04±0.03 to 0.05±0.12 $mgC \cdot m^{-2} \cdot h^{-1}$ for 2016. The considerable increase of the methane concentration within the lower part of the profile indicated high methanogenic activity. It was not observed at the upper layer of the soil because the produced methane was consumed by methanotrophs. We suppose that changing environmental conditions (e.g., moisture increase or temperature decrease) may result in methanotrophic activity decrease. In this case, forest sites with the above mentioned non-monotonic concentration profiles may likely become methane sources instead of sinks.

For example, we found $CH_4$ consumption at the Tr.PWF site in 2016 when the WTL was low. On the contrary, $CH_4$ emission was found at this site in 2017 when the WTL was higher. The changes in the soil concentration profiles are shown in the figure 3.b. In 2016, the $CH_4$ concentration was lowest at 15 cm depth that indicated high activity of methanotrophs consuming methane from upper and lower soil layers. When the WTL increased in 2017, the methanotrophic activity decreased so the $CH_4$ concentrations were constant at 5-15 cm depths. As a result, produced $CH_4$ was not completely consumed by methanotrophs enabling it to emit into the atmosphere.

The $CH_4$ flux measurements from soils at Wet Forest_Polynyanka and Wet Forest_3 sites with different WTL showed that $CH_4$ emission shifted to $CH_4$ consumption when the WTL was about 15-45 cm. More measurements are needed to understand the environmental controls of this shift.

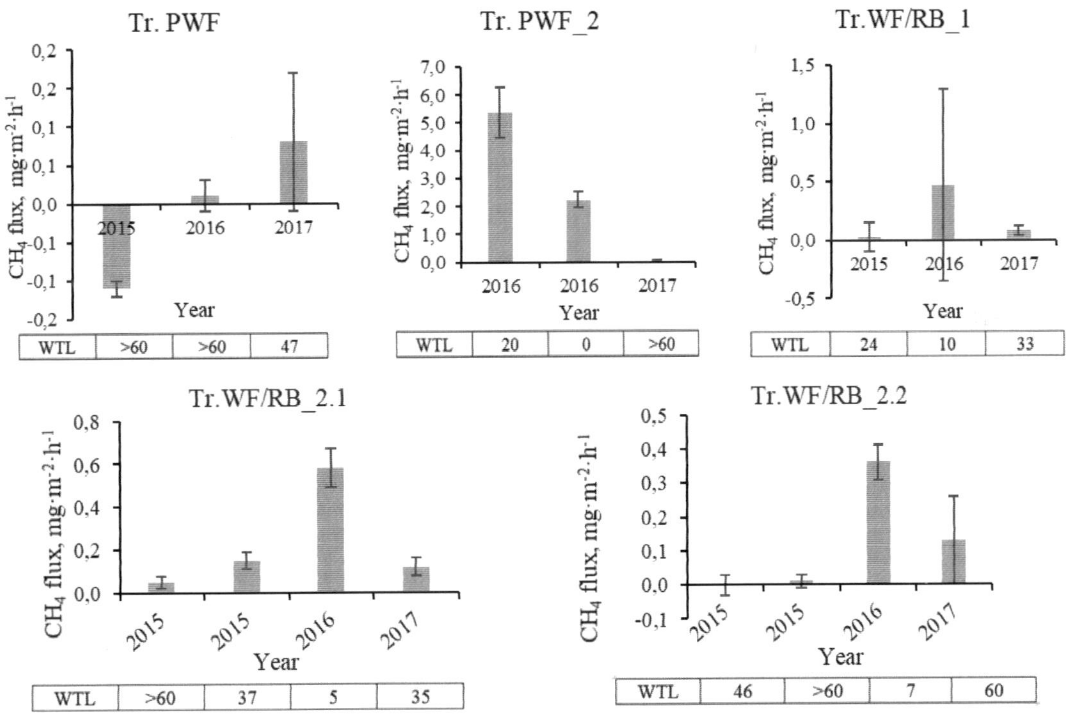


**Figure 4.** Methane fluxes and WTL at Wet Forest, Wet Forest_2, WF/RB sites in 2015-2017.

The methane flux measurements made in 2015-2017 revealed the strong influence of the WTL on methane flux at Tr.WF/RB_2.1 and Tr.WF/RB_2.2 sites (figure 4). However, the measurement error was too high to reliably determine WTL impact on methane emission at the Tr. PWF and Tr.WF/RB_1 sites. Measurements at the Tr. PWF_2 site showed that methane flux was higher at the local elevation (WTL of 20 cm) than at the local depression (WTL of 0 cm) in 2016. This may indicate another factor influencing the methane flux.

Values of carbon dioxide fluxes from soil were generally consistent with result of researches which were conducted in flooded forests of south taiga during the summer season earlier [16]. Also [16] shows influence of WTL on intensity of $CO_2$ emission, but dynamics of change in fluxes with WTL was different from our results (figure 2.b, figure 2.d). Perhaps reasons of this dissonance are differences in conditions of measurements, such as composition of grass cover.

Our values of carbon dioxide emission in south taiga during summer season exceed annual average for this zone [17, 18]. Such exceedance is confirmed by [19].

## 5. Conclusion

We may conclude that the seasonally flooded forests may produce methane when the WTL is higher than 15-45 cm when the methanotrophic activity decreases. The measurements at the soil surface do not reveal any methane fluxes from the soil at the sites with deeper WTL during the measurement period. But these sites may become potential methane sources if the methane concentration profile shows the methane concentration increase at a certain depth. This increase may be an evidence of the methanogenic activity that may lead to methane emission when the methanotrophs layer will become less active or not active at all (it may happen for example after a heavy rain).

**Acknowledgements**

The study was supported by ArCS - Arctic Challenge for Sustainability Project. The authors express sincere gratitude to M.V. Chistotin (All-Russian Scientific and Research Institute of Agrochemistry named by D.N. Pryanishnikov) for providing the analiser DX6100-01, the specialists from V.E. Zuev Institute of Atmospheric Optics D.K. Davydov, A.V. Fofonov and M.Yu. Arshinov, who calibrated the instrument, I.S. Churkin for assistance with translation, A.A. Lebedev, who created the software, which allowed us to make the calculations of the average temperature and the weighted average much faster as well as M.V. Lukin and G.V. Uvarov for participating in the field measurements.

**Appendix.** Summary of individual measured methane and carbon dioxide fluxes, 2017.

| Point | Date (of 2017) | $CH_4$ flux, mg $CH_4 \cdot m^{-2} \cdot h^{-1}$ | | | $CO_2$ flux, mg $C \cdot m^{-2} \cdot h^{-1}$ | | | Temperature $CH_4/CO_2$, °C | | | | | WTL, cm |
|---|---|---|---|---|---|---|---|---|---|---|---|---|---|
| | | Mean | Error | Error code[a)] | Mean | Error | Error code [a)] | of air | of soil at depth, cm: 0 | 5 | 15 | 45 | |
| T.Plo.H.1.3 | 27.07 | 31.101 | 3.275 | 6 | - | - | - | 29,3/- | 24,7/- | 23,6/- | 19,5/- | 14,0/-[b)] | 0 |
| T.Plo.H.1.3 | 27.07 | 26.268 | 6.013 | 4 | - | - | - | 32,2/- | 29,3/- | 24,5/- | 19,5/- | 14,0/-[b)] | 0 |
| T.Plo.H.1.3 | 27.07 | 33.369 | 5.507 | 4 | 416 | 62 | 4 | 31.9/33.0 | 30.6/31.3 | 24.5/24.5 | 19.1/19.0 | 14.0/14.0 [b)] | 0 |
| fp_2 | 01.08 | 9.231 | 1.024 | 6 | 54 | 112 | 4 | 28.9/29.5 | 29.8/25.8 | 23.6/22.4 | 15.0/15.0 | 12.0/12.0 | 0 |
| fp_2 | 01.08 | 9.687 | 1.563 | 4 | 17 | 7 | 4 | 30.2/29.0 | 29.4/30.8 | 24.6/25.0 | 15.6/16.4 | 12.1/12.5 | 0 |
| fp_2 | 01.08 | 8.133 | 4.035 | 4 | 0 | 2 | 95 | 27.0/27.7 | 23.2/24.8 | 20.8/21.6 | 15.0/15.0 | 12.0/12.0 | 0 |
| fp_3 | 01.08 | 18.524 | 1.915 | 4 | 370 | 54 | 4 | 30.3/29.2 | 23.6/23.0 | 21.0/20.5 | 19.5/19.4 | 14.1/14.4 | -15 |
| fp_3 | 01.08 | 17.745 | 1.392 | 4 | 209 | 23 | 4 | 28.4/29.4 | 23.4/23.2 | 21.7/21.3 | 20.0/19.9 | 13.6/13.8 | -15 |
| fp_3 | 01.08 | 21.073 | 1.959 | 6 | 77 | 74 | 4 | 26.2/27.3 | 22.9/23.5 | 21.9/22.0 | 19.8/20.0 | 13.5/13.5 | -15 |
| fp_3 | 01.08 | 18.944 | 6.015 | 4 | 5 | 112 | 4 | 26.8/27.5 | 23.2/23.5 | 22.0/22.0 | 19.8/20.0 | 13.5/13.5 | -15 |
| fp_3 | 01.08 | 31.268 | 3.191 | 4 | 17 | 79 | 4 | 28.9/29.3 | 23.4/23.1 | 21.5/21.3 | 20.0/20.0 | 13.6/13.8 | -15 |
| fp_3 | 01.08 | 21.817 | 2.996 | 6 | 75 | 16 | 4 | 30.2/29.1 | 23.6/23.1 | 20.9/20.5 | 19.5/19.3 | 14.1/14.5 | -15 |
| fp_5 | 15.08 | 1.027 | 0.265 | 6 | - | - | - | 15.0/ - - | 12.0/ - - | 12.5/ - - | 12.0/ - - | 13.0/ - - | 2.5 |
| fp_5 | 15.08 | 0.697 | 0.117 | 6 | 142 | 112 | 4 | 14.9/14.9 | 12.0/12.0 | 12.5/12.8 | 12.0/12.0 | 13.0/13.0 | 2.5 |
| fp_5 | 15.08 | 0.720 | 0.100 | 4 | 0 | 5 | 95 | 15.7/15.0 | 12.0/12.0 | 12.5/12.5 | 12.0/12.0 | 12.8/12.9 | 2.5 |
| fp_5 | 15.08 | 0.636 | 0.115 | 6 | 30 | 86 | 4 | 15.2/15.0 | 12.0/12.0 | 12.5/12.5 | 12.0/12.0 | 12.8/13.0 | 2.5 |
| fp_5 | 15.08 | 0.208 | 0.156 | 4 | 88 | 142 | 4 | 17.3/17.0 | 12.5/12.5 | 12.5/12.5 | 12.0/12.0 | 12.6/12.6 | 2.5 |
| fp_5 | 15.08 | 0.664 | 0.121 | 6 | 93 | 127 | 4 | 17.0/16.5 | 12.3/12.0 | 12.5/12.5 | 12.0/12.0 | 12.7/12.8 | 2.5 |
| fp_5 | 15.08 | 0.568 | 0.081 | 4 | - | - | - - | 15.3/ - - | 12.0/ - - | 12.5/ - - | 12.0/ - - | 12.8/ - - | 2.5 |
| fp_5 | 15.08 | 0.581 | 0.256 | 6 | 64 | 162 | 4 | 16.8/16.2 | 12.3/12.0 | 12.5/12.5 | 12.0/12.0 | 12.7/12.8 | 2.5 |
| fp_5 | 15.08 | 0.723 | 0.145 | 6 | 142 | 112 | 4 | 14.9/14.9 | 12.0/12.0 | 12.5/12.8 | 12.0/12.0 | 13.0/13.0 | 2.5 |
| fp_6 | 09.08 | 0.145 | 0.179 | 6 | - | - | - | 18,9/- | 16,0/- | 15,5/- | 15,2/- | 14,0/- | 5 |
| fp_6 | 09.08 | 0.075 | 0.132 | 6 | - | - | - | 18,1/- | 16,0/- | 15,5/- | 15,0/- | 14,0/- | 5 |
| fp_6 | 09.08 | -0.044 | 0.260 | 95 | 483 | 45 | 4 | 15.3/16.5 | 16.0/16.0 | 15.5/15.5 | 15.0/15.0 | 14.0/14.0 | 5 |
| fp_6 | 09.08 | -0.173 | 0.359 | 95 | 113 | 202 | 4 | 14.5/14.5 | 15.5/15.5 | 15.5/15.5 | 15.0/15.0 | 14.0/14.0 | 5 |
| fp_7 | 09.08 | 0.024 | 0.040 | 4 | - | - | - | 18,8/- | 16,7/- | 14,5/- | 14,5/- | 13,5/- | 15 |
| fp_7 | 09.08 | -0.016 | 0.075 | 95 | - | - | - | 17,5/- | 16,5/- | 15,0/- | 14,5/- | 13,5/- | 15 |
| fp_7 | 09.08 | -0.037 | 0.016 | 4 | - | - | - | 14,7/- | 16,2/- | 15,0/- | 14,5/- | 13,5/- | 15 |
| fp_7 | 09.08 | 0.021 | 0.172 | 6 | - | - | - | 14,7/- | 16,0/- | 15,0/- | 14,5/- | 13,5/- | 15 |
| fp_8 | 15.08 | 0.025 | 0.065 | 6 | 323 | 431 | 4 | 15.2/15.1 | 14.3/14.8 | 12.6/12.8 | 11.6/11.9 | 11.4/11.5 | 45 |

| | | | | | | | | | | | | | |
|---|---|---|---|---|---|---|---|---|---|---|---|---|---|
| fp_8 | 15.08 | -0.023 | 0.111 | 95 | 373 | 104 | 4 | 15.2/15.0 | 14.0/11.5 | 12.0/12.0 | 11.5/11.5 | 11.0/11.1 | 45 |
| fp_8 | 15.08 | -0.048 | 0.179 | 95 | 115 | 243 | 4 | 16.7/16.0 | 14.6/14.1 | 12.3/12.0 | 11.5/11.5 | 11.0/11.0 | 45 |
| fp_8 | 09.08 | -0.045 | 0.113 | 95 | - | - | - | 18,2/- | 16,4/- | 15,0/- | 13,0/- | 12,0/- | 45 |
| fp_8 | 09.08 | -0.041 | 0.581 | 95 | - | - | - | 18,9/- | 18,0/- | 16,8/- | 14,7/ | 12,6/- | 45 |
| fp_8 | 09.08 | -0.039 | 0.080 | 95 | - | - | - | 16,3/- | 15,8/- | 15,0/- | 13,0/- | 12,0/- | 45 |
| fp_8 | 09.08 | -0.024 | 0.065 | 95 | - | - | - | 14,5/- | 15,3/- | 15,0/- | 13,0/- | 12,0/- | 45 |
| fp_9 | 09.08 | -2.513 | 3.241 | 95 | - | - | - | 19,0/- | 19,9/- | 17,1/- | 14,3/-[c)] | -/- | 75 |
| fp_9 | 09.08 | -0.056 | 0.080 | 95 | - | - | - | 18,2/- | 19,1/- | 16,5/- | 13,2/-[c)] | -/- | 75 |
| fp_9 | 09.08 | -0.031 | 0.400 | 95 | - | - | - | 16,3/- | 17,9/- | 16,3/- | 13,0/-[c)] | -/- | 75 |
| fp_9 | 09.08 | -0.161 | 0.131 | 95 | - | - | - | 14,5/- | 16,5/ | 15,9/- | 13,0/-[c)] | -/- | 75 |
| WF_BD_01 | 04.08 | -0.017 | 1.577 | 95 | 66 | 89 | 4 | 31.2/30.2 | 28.4/36.5 | 22.9/29.3 | 19.0/21.6 | 16.9/18.2 | 0 |
| WF_BD_01 | 04.08 | -0.179 | 0.465 | 95 | 22 | 104 | 4 | 31.3/31.3 | 25.4/25.5 | 20.9/20.5 | 18.0/18.0 | 16.5/16.5 | 0 |
| WF_BD_02 | 04.08 | 0.352 | 0.565 | 4 | 130 | 9 | 4 | 31.5/29.3 | 29.3/31.8 | 23.2/25.9 | 18.9/21.4 | 17.4/19.3 | 10 |
| WF_BD_02 | 04.08 | 0.273 | 1.180 | 6 | 70 | 41 | 4 | 31.2/31.0 | 29.1/29.0 | 22.0/22.0 | 18.0/18.0 | 16.8/17.0 | 10 |
| WF_BD_1 | 06.08 | 20.172 | 5.443 | 4 | 25 | 545 | 4 | 26.5/27.3 | 23.7/24.1 | 19.0/19.1 | 17.0/17.0 | 14.6/14.7 | 0 |
| WF_BD_1 | 06.08 | 5.417 | 0.243 | 4 | - | - | - | 28,1/- | 24,3/- | 19,0/- | 16,5/- | 14,5/- | 0 |
| WF_BD_1 | 06.08 | 7.199 | 2.013 | 4 | 0 | 2 | 95 | 28.3/28.6 | 22.7/23.2 | 19.5/19.5 | 17.0/17.0 | 14.5/14.5 | 0 |
| WF_BD_1 | 06.08 | 4.243 | 0.979 | 6 | - - | - - | - - | 28.9/ - - | 23.4/ - - | 19.5/ - - | 17.0/ - - | 14.5/ - - | 0 |
| WF_BD_1 | 06.08 | 5.693 | 0.488 | 6 | 281 | 21 | 4 | 27.4/27.5 | 23.9/24.0 | 19.0/19.0 | 16.7/16.5 | 14.5/14.5 | 0 |
| WF_BD_2 | 17.08 | 2.223 | 0.396 | 4 | 143 | 65 | 4 | 17.4/19.0 | 16.8/18.1 | 14.6/15.0 | 14.0/14.0 | 13.1/13.5 | 2 |
| WF_BD_2 | 19.08 | 0.987 | 0.237 | 4 | 454 | 286 | 4 | 17.6/17.5 | 16.3/16.2 | 16.0/15.9 | 15.8/15.7 | 14.4/14.3 | 2 |
| WF_BD_2 | 19.08 | 0.729 | 0.277 | 6 | 138 | 157 | 4 | 19.1/17.7 | 16.8/16.3 | 15.5/16.1 | 14.0/15.9 | 13.0/14.5 | 2 |
| WF_BD_2 | 19.08 | 0.709 | 0.295 | 4 | 220 | 568 | 4 | 17.1/17.9 | 15.6/16.5 | 15.0/16.4 | 14.0/16.2 | 13.0/14.7 | 2 |
| WF_BD_2 | 19.08 | 0.392 | 0.036 | 4 | - - | - - | - - | 16.6/ - - | 15.2/ - - | 15.0/ - - | 14.0/ - - | 13.0/ - - | 2 |
| WF_BD_2 | 19.08 | 0.344 | 0.172 | 6 | 133 | 218 | 4 | 14.5/18.4 | 14.6/16.9 | 15.0/16.8 | 14.0/16.7 | 12.9/15.2 | 2 |
| WF_BD_2 | 19.08 | 9.108 | 1.251 | 4 | 153 | 72 | 4 | 19.8/17.1 | 17.4/15.7 | 14.5/15.2 | 13.6/14.9 | 13.0/13.7 | 2 |
| WF_BD_2 | 19.08 | 2.561 | 0.717 | 4 | 316 | 453 | 4 | 20.8/17.2 | 17.8/15.9 | 14.9/15.4 | 13.6/15.1 | 13.0/13.9 | 2 |
| WF_BD_2 | 19.08 | 2.317 | 0.428 | 4 | 184 | 121 | 4 | 20.1/17.3 | 17.5/16.0 | 15.0/15.6 | 14.0/15.4 | 13.0/14.1 | 2 |
| WF_BD_3 | 17.08 | 10.541 | 5.561 | 4 | 0 | 13 | 95 | 18.3/21.2 | 17.0/19.5 | 15.5/17.2 | 14.0/14.7 | 13.5/13.8 | 7 |
| WF_BD_3 | 17.08 | -0.025 | 0.880 | 95 | 54 | 220 | 4 | 17.4/17.1 | 15.6/15.8 | 15.0/15.0 | 13.5/13.5 | 13.5/13.5 | 7 |
| WF_BD_3 | 19.08 | -0.059 | 0.267 | 95 | 71 | 14 | 4 | 18.9/15.5 | 16.9/15.0 | 16.5/14.0 | 16.5/14.0 | 14.6/13.0 | 7 |
| WF_BD_3 | 19.08 | 0.192 | 0.048 | 4 | 66 | 119 | 4 | 18.6/17.3 | 16.6/15.0 | 16.2/14.0 | 16.2/14.0 | 14.3/13.0 | 7 |
| WF_BD_3 | 19.08 | 0.239 | 0.091 | 6 | 94 | 58 | 4 | 18.3/19.6 | 16.3/15.5 | 15.8/14.0 | 15.8/14.0 | 14.1/13.0 | 7 |
| WF_BD_3 | 19.08 | 0.939 | 0.309 | 4 | 19 | 126 | 4 | 17.6/17.4 | 15.4/15.3 | 14.9/14.7 | 14.9/14.7 | 13.5/13.4 | 7 |
| WF_BD_3 | 19.08 | 0.511 | 0.123 | 4 | - - | - - | - - | 17.8/ - - | 15.7/ - - | 15.2/ - - | 15.2/ - - | 13.7/ - - | 7 |
| WF_BD_3 | 19.08 | 0.596 | 0.304 | 4 | 59 | 66 | 4 | 18.0/20.1 | 15.9/15.5 | 15.5/14.0 | 15.5/14.0 | 13.8/13.0 | 7 |
| WF_BD_4 | 04.08 | 2.165 | 0.772 | 4 | 67 | 8 | 4 | 31.1/32.0 | 27.8/27.0 | 21.2/20.5 | 16.5/15.8 | 13.6/13.5 | 10 |
| WF_BD_5 | 06.08 | 0.072 | 0.181 | 6 | - | - | - | 28,8/- | 19,9/- | 17,0/- | 15,0/- | 13,0/- | 15 |
| WF_BD_5 | 06.08 | 7.696 | 0.984 | 4 | 396 | 610 | 4 | 25.9/26.3 | 22.6/24.4 | 18.3/18.9 | 16.5/17.2 | 13.7/14.0 | 15 |
| WF_BD_5 | 06.08 | 0.816 | 0.575 | 4 | 0 | 2 | 95 | 28.1/27.4 | 22.0/22.0 | 18.0/18.0 | 16.0/16.0 | 13.5/13.5 | 15 |
| WF_BD_5 | 06.08 | 0.085 | 0.199 | 6 | 175 | 194 | 4 | 29.2/30.3 | 19.5/19.5 | 17.0/17.0 | 15.0/15.0 | 13.0/13.0 | 15 |
| WF_BD_5 | 06.08 | 0.175 | 0.191 | 6 | - | - | - | 27,8/- | 20,3/- | 17,2/- | 15,0/- | 13,0/- | 15 |
| WF_BD_5 | 06.08 | 2.388 | 0.580 | 4 | 155 | 134 | 4 | 27.8/27.6 | 22.2/22.0 | 18.0/18.0 | 16.0/16.0 | 13.5/13.5 | 15 |
| WF_BD_5 | 06.08 | 0.515 | 0.200 | 6 | 233 | 12 | 6 | 28.9/29.3 | 22.0/22.0 | 18.0/18.0 | 16.0/16.0 | 13.5/13.5 | 15 |
| WF_BD_5 | 06.08 | 0.588 | 0.144 | 6 | 305 | 179 | 4 | 27.6/28.6 | 21.8/22.0 | 18.0/18.0 | 16.0/16.0 | 13.5/13.5 | 15 |
| WF_BD_5 | 06.08 | 0.025 | 0.080 | 6 | - | - | - | 27,3/- | 19,5/- | 17,0/- | 15,0/- | 13,0/- | 15 |
| WF_BD_5 | 06.08 | 0.747 | 0.668 | 4 | - | - | - | 26.4/ - - | 21.0/ - - | 18.1/ - - | 15.8/ - - | 13.1/ - - | 15 |
| WF_BD_6 | 04.08 | 4.836 | 0.361 | 4 | 345 | 54 | 4 | 31.3/31.1 | 27.9/30.6 | 21.2/22.0 | 16.5/17.8 | 13.6/14.0 | 20 |
| WF_BD_7 | 06.08 | 0.059 | 0.077 | 6 | - | - | - | 28,0/- | 20,4/- | 17,5/- | 14,0/- | 11,0/- | 30 |
| WF_BD_7 | 06.08 | 0.011 | 0.163 | 6 | - | - | - | 28,1/- | 20,5/- | 17,5/- | 14,0/- | 11,0/- | 30 |
| WF_BD_7 | 06.08 | 0.457 | 0.115 | 4 | 351 | 30 | 4 | 26.1/26.6 | 20.0/20.8 | 17.4/18.3 | 14.6/15.0 | 11.1/11.5 | 30 |
| WF_BD_7 | 06.08 | 0.352 | 0.969 | 4 | - | - | - | 26,3/- | 20,4/- | 17,9/- | 14,8/- | 11,3/- | 30 |
| WF_BD_7 | 06.08 | -0.001 | 0.065 | 95 | - - | - - | - - | 29.7/ - - | 20.1/ - - | 17.0/ - - | 14.0/ - - | 11.0/ - - | 30 |
| WF_BD_7 | 06.08 | -0.028 | 0.048 | 95 | 5 | 245 | 4 | 29.0/28.2 | 20.0/20.0 | 16.7/16.5 | 14.0/14.0 | 11.0/11.0 | 30 |
| WF_BD_7 | 06.08 | 0.071 | 0.121 | 6 | 485 | 1399 | 4 | 29.4/28.5 | 20.0/20.0 | 16.8/16.5 | 14.0/14.0 | 11.0/11.0 | 30 |
| WF_BD_7 | 06.08 | 0.189 | 0.149 | 6 | 225 | 35 | 4 | 28.8/28.3 | 20.3/20.1 | 16.9/17.0 | 14.3/14.5 | 11.0/11.0 | 30 |
| WF_BD_7 | 06.08 | 0.057 | 0.121 | 4 | 237 | 339 | 4 | 29.0/28.3 | 20.5/20.0 | 17.0/17.0 | 14.4/14.5 | 11.0/11.0 | 30 |
| WF_BD_7 | 06.08 | -0.012 | 0.245 | 95 | - | - | - | 29,5/- | 20,1/- | 17,0/- | 14,0/- | 11,0/- | 30 |
| WF_BD_8 | 17.08 | 0.028 | 0.072 | 6 | - | - | - | 17,1/- | 18,0/- | 16,7/- | 14,8/- | 11,3/- | 45 |
| WF_BD_8 | 17.08 | -0.061 | 0.139 | 95 | 506 | 108 | 4 | 17.2/17.0 | 17.5/18.1 | 16.0/16.7 | 14.2/14.8 | 11.1/11.5 | 45 |
| WF_BD_8 | 17.08 | -0.001 | 0.684 | 95 | - | - | - | 17,5/- | 17,0/- | 15,3/- | 13,6/- | 11,0/- | 45 |
| WF_BD_8 | 17.08 | - | - | - | 299 | 211 | 4 | - /17.5 | - /17.0 | - /15.3 | - /13.5 | - /11.0 | 45 |
| WF_BD_8 | 19.08 | -0.012 | 0.051 | 95 | 127 | 148 | 4 | 17.1/17.4 | 14.7/15.0 | 15.1/14.7 | 14.1/13.5 | 13.0/11.5 | 45 |
| WF_BD_8 | 19.08 | -0.008 | 0.173 | 95 | 128 | 48 | 4 | 17.1/17.8 | 14.8/15.0 | 15.1/14.6 | 14.1/13.5 | 13.0/11.5 | 45 |
| WF_BD_8 | 19.08 | -0.049 | 0.033 | 95 | - | - | - | 17,1/- | 14,9/- | 15,4/- | 14,3/- | 13,3/- | 45 |

| | | | | | | | | | | | | | |
|---|---|---|---|---|---|---|---|---|---|---|---|---|---|
| WF_BD_8 | 19.08 | -0.024 | 0.072 | 95 | 303 | 112 | 4 | 17.1/17.8 | 15.0/15.1 | 15.4/14.5 | 14.4/13.0 | 13.3/11.0 | 45 |
| WF_BD_8 | 19.08 | -0.025 | 0.181 | 95 | 322 | 127 | 4 | 17.1/19.9 | 15.2/15.5 | 15.7/15.0 | 14.7/12.5 | 13.6/11.0 | 45 |
| WF_BD_8 | 19.08 | -0.023 | 0.037 | 95 | 250 | 88 | 4 | 17.1/20.2 | 15.2/15.5 | 15.7/15.0 | 14.7/12.5 | 13.6/11.0 | 45 |
| WF_BD_8 | 19.08 | -0.019 | 0.087 | 95 | 155 | 97 | 4 | 17.0/20.8 | 15.4/15.7 | 15.9/15.0 | 15.0/13.0 | 13.6/11.0 | 45 |
| WF_BD_8 | 19.08 | -0.080 | 0.049 | 95 | 15 | 19 | 4 | 17.0/20.6 | 15.4/15.5 | 16.0/15.0 | 15.0/13.0 | 13.9/11.0 | 45 |
| WF_BD_8 | 19.08 | 0.000 | 0.076 | 4 | 330 | 130 | 4 | 16.9/19.8 | 15.7/15.5 | 16.2/15.0 | 15.3/13.0 | 14.2/11.0 | 45 |
| WF_BD_8 | 19.08 | 0.048 | 0.104 | 6 | 191 | 76 | 4 | 16.9/19.8 | 15.7/15.4 | 16.3/15.0 | 15.4/13.0 | 14.3/11.0 | 45 |
| WF_BD_8 | 19.08 | -0.023 | 0.129 | 95 | 205 | 155 | 4 | 16.9/17.0 | 16.2/14.5 | 16.8/14.5 | 16.0/13.0 | 15.0/11.0 | 45 |
| WF_BD_8 | 19.08 | -0.029 | 0.200 | 95 | 390 | 30 | 4 | 16.9/16.9 | 15.9/14.5 | 16.5/14.9 | 15.7/13.0 | 14.6/11.0 | 45 |
| WF_BD_8 | 19.08 | -0.055 | 0.121 | 95 | 172 | 254 | 4 | 16.9/16.7 | 15.9/14.5 | 16.6/14.7 | 15.7/13.0 | 14.6/11.0 | 45 |
| WF_BD_8 | 19.08 | -0.077 | 0.147 | 95 | 101 | 82 | 4 | 16.9/17.0 | 16.2/14.3 | 16.9/14.5 | 16.1/13.0 | 15.1/11.0 | 45 |
| Tr. PWF | 21.08 | 0.269 | 0.121 | 4 | 1 | 5 | 95 | 18.3/19.4 | 15.0/15.0 | 14.0/14.0 | 13.6/14.0 | 12.0/12.0 | 47 |
| Tr. PWF | 21.08 | 0.271 | 0.057 | 4 | 3 | 2 | 95 | 18.3/19.6 | 15.0/15.0 | 14.0/14.0 | 13.6/14.0 | 12.0/12.0 | 47 |
| Tr. PWF | 21.08 | 0.056 | 0.028 | 4 | - | - | - | 17,6/- | 14,1/- | 13,5/- | 12,5/- | 12,1/- | 47 |
| Tr. PWF | 21.08 | 0.023 | 0.039 | 4 | - | - | - | 17,6/- | 14,1/- | 13,5/- | 12,5/- | 12,1/- | 47 |
| Tr. PWF_2 | 21.08 | 0.007 | 0.060 | 6 | 397 | 217 | 4 | 17.1/17.0 | 16.4/16.5 | 16.1/16.5 | 15.1/15.9 | 16.5/17.5 | - [d] |
| Tr. PWF_2 | 21.08 | 0.068 | 0.063 | 6 | 196 | 26 | 4 | 17.1/17.3 | 16.2/16.5 | 15.7/16.5 | 14.6/15.6 | 15.9/17.5 | - [d] |
| Tr. PWF_2 | 21.08 | 0.115 | 0.116 | 6 | - | - | - | 17,5/- | 14,3/- | 13,5/- | 12,5/- | 12,3/- | - [d] |
| Tr. PWF_2 | 21.08 | 0.036 | 0.219 | 6 | 515 | 194 | 4 | 17.5/17.5 | 14.2/14.5 | 13.5/13.5 | 12.5/12.5 | 12.2/12.5 | - [d] |
| Tr.WF/RB_1 | 16.08 | 0.101 | 0.091 | 4 | 9 | 64 | 4 | 20.9/21.0 | 17.5/17.5 | 15.7/16.0 | 13.0/13.0 | 11.6/11.8 | 30 |
| Tr.WF/RB_1 | 16.08 | 0.000 | 0.109 | 6 | - - | - - | - - | 20.2/ - - | 16.7/ - - | 15.5/ - - | 13.0/ - - | 11.5/ - - | 30 |
| Tr.WF/RB_1 | 16.08 | 0.064 | 0.185 | 6 | 272 | 118 | 4 | 20.1/20.5 | 19.5/20.0 | 18.2/18.5 | 14.0/14.0 | 12.0/12.0 | 35 |
| Tr.WF/RB_1 | 16.08 | 0.089 | 0.048 | 4 | 360 | 141 | 4 | 20.7/21.0 | 20.7/21.0 | 18.9/19.0 | 13.8/13.6 | 12.0/12.0 | 35 |
| Tr.WF/RB_2.2 | 16.08 | 0.304 | 0.084 | 6 | 411 | 202 | 4 | 19.8/20.1 | 19.3/20.1 | 15.5/16.0 | 13.6/14.2 | 12.7/13.2 | 60 |
| Tr.WF/RB_2.2 | 16.08 | 0.052 | 0.059 | 4 | 422 | 238 | 4 | 20.5/19.8 | 17.8/19.8 | 15.0/15.5 | 13.0/13.5 | 12.5/12.5 | 60 |
| Tr.WF/RB_2.1 | 16.08 | 0.153 | 0.067 | 4 | 0 | 2 | 95 | 20.1/20.0 | 16.5/17.6 | 14.9/15.7 | 13.3/13.7 | 12.2/12.5 | 35 |
| Tr.WF/RB_2.1 | 16.08 | 0.099 | 0.055 | 6 | 107 | 94 | 4 | 20.5/19.8 | 16.0/16.5 | 14.5/15.0 | 13.0/13.5 | 12.0/12.5 | 35 |
| Ryam_1 | 10.08 | -0.023 | 0.291 | 95 | 2 | 3 | 95 | 20.9/21.3 | 18.6/18.5 | 16.1/16.3 | 14.4/14.5 | 11.9/12.0 | 102 |
| Ryam_1 | 10.08 | -0.213 | 0.617 | 95 | 1 | 6 | 95 | 20.3/20.8 | 18.9/18.7 | 16.3/16.0 | 14.1/14.2 | 11.6/11.8 | 102 |
| Ryam_1 | 10.08 | 0.064 | 0.081 | 6 | 2 | 2 | 95 | 20.5/20.0 | 19.2/19.0 | 16.6/16.5 | 14.3/14.0 | 11.5/11.6 | 102 |
| Ryam_1 | 10.08 | -0.489 | 0.728 | 95 | - - | - - | - - | 21.1/ - - | 20.4/ - - | 17.1/ - - | 14.5/ - - | 11.5/ - - | 102 |
| Ryam_1 | 10.08 | -0.027 | 0.044 | 95 | 1 | 4 | 95 | 20.0/21.9 | 20.6/20.6 | 17.4/17.0 | 14.5/14.5 | 11.5/11.5 | 102 |
| R_1 | 13.08 | 0.067 | 0.129 | 6 | - | - | - | 18,8/- | 17,2/- | 15,9/- | 13,7/- | 10,3/- | 43 |
| R_1 | 13.08 | -0.047 | 0.331 | 95 | 114 | 232 | 4 | 18.9/19.3 | 17.3/17.0 | 15.5/15.6 | 13.1/13.5 | 10.1/10.3 | 43 |
| R_1 | 13.08 | -0.035 | 0.081 | 95 | 0 | 5 | 95 | 18.4/18.2 | 17.1/17.4 | 15.5/15.5 | 13.0/13.0 | 10.0/10.0 | 43 |
| R_1 | 13.08 | -0.019 | 0.019 | 95 | 73 | 108 | 4 | 18.7/18.5 | 17.3/17.0 | 15.5/15.5 | 13.0/13.0 | 10.0/10.0 | 43 |
| R_1 | 13.08 | -0.008 | 0.151 | 95 | 172 | 197 | 4 | 19.2/19.1 | 17.6/17.7 | 15.8/15.5 | 13.0/13.0 | 10.0/10.0 | 43 |
| R_1 | 13.08 | -0.064 | 0.128 | 95 | 134 | 31 | 4 | 19.0/19.1 | 17.4/17.5 | 16.0/16.0 | 13.0/13.0 | 10.0/10.0 | 43 |
| R_1 | 13.08 | 0.031 | 0.044 | 6 | - | - | - | 18,3/- | 16,6/- | 15,9/- | 13,0/- | 10,0/- | 43 |
| R_1 | 13.08 | 0.067 | 0.036 | 6 | 181 | 62 | 4 | 17.6/18.1 | 16.0/16.6 | 15.7/16.0 | 13.1/13.0 | 10.0/10.0 | 43 |
| R_1 | 13.08 | 0.027 | 0.085 | 6 | 380 | 48 | 4 | 15.7/15.5 | 14.9/15.5 | 15.4/15.5 | 13.4/13.2 | 10.0/10.0 | 43 |
| R_1 | 13.08 | -0.035 | 0.225 | 95 | 0 | 3 | 95 | 15.4/15.0 | 14.7/15.0 | 15.4/15.5 | 13.5/13.5 | 10.0/10.0 | 43 |
| R_2 | 13.08 | -0.012 | 0.145 | 95 | 349 | 101 | 4 | 18.5/17.5 | 18.8/18.7 | 17.4/18.2 | 14.3/15.1 | 10.2/10.5 | 65 |
| R_2 | 13.08 | 0.317 | 0.092 | 4 | 19 | 143 | 4 | 17.9/18.3 | 18.2/18.7 | 16.5/16.5 | 13.5/13.5 | 10.0/10.0 | 65 |
| R_2 | 13.08 | 0.099 | 0.133 | 6 | 37 | 152 | 4 | 18.5/18.4 | 18.2/18.1 | 16.5/16.5 | 13.5/13.5 | 10.0/10.0 | 65 |
| R_2 | 13.08 | 0.047 | 0.071 | 4 | 231 | 211 | 4 | 17.9/18.1 | 17.9/18.0 | 16.5/16.5 | 13.5/13.5 | 10.0/10.0 | 65 |
| R_2 | 13.08 | 0.072 | 0.133 | 6 | 153 | 19 | 4 | 16.5/17.3 | 16.9/17.5 | 16.1/16.3 | 13.5/13.5 | 10.0/10.0 | 65 |
| R_3 | 13.08 | 0.008 | 0.064 | 6 | 164 | 168 | 4 | 18.5/17.5 | 18.8/18.8 | 17.5/18.3 | 14.4/15.3 | 10.2/10.5 | 65 |
| R_3 | 13.08 | 0.055 | 0.141 | 6 | 57 | 64 | 4 | 17.9/18.5 | 18.3/18.8 | 16.5/16.6 | 13.5/13.5 | 10.0/10.0 | 65 |
| R_3 | 13.08 | -0.013 | 0.084 | 95 | 86 | 146 | 4 | 18.5/18.3 | 18.2/18.0 | 16.5/16.5 | 13.5/13.5 | 10.0/10.0 | 65 |
| R_3 | 13.08 | 0.033 | 0.112 | 6 | 193 | 139 | 4 | 17.8/18.1 | 17.9/18.0 | 16.5/16.5 | 13.5/13.5 | 10.0/10.0 | 65 |
| R_3 | 13.08 | 0.072 | 0.089 | 4 | 254 | 50 | 4 | 16.3/17.3 | 16.7/17.5 | 16.0/16.1 | 13.5/13.4 | 10.0/10.0 | 65 |

[a] error code: 4 - standard error, 6 - error derived from solving linear programming problem, 95 - confidence interval is 95%;
[b] measurements at 55 cm depth;
[c] measurements at 25 cm depth;
[d] WTL was deeper than 60 cm. The water content was 61.1±3.3, 63.6±4.6 and 78.8±5.1% at soil depth of 5, 10 and 15 cm respectively.

## References

[1] Andronova N G and Karol I L 1993 The contribution of USSR sources to global methane emission *Chemosphere* **26** 111–26

[2] Alm J, Korhola A, Turunen J, Saarnio S, Jungner H, Tolonen K and Silvola J 1999 Past and future atmospheric carbon gas ($CO_2$, $CH_4$) exchange in boreal peatlands *Int. Peat J.* **9** 127–35
[3] Gal'chenko V F, Dulov L E, Cramer B, Konova N I and Barysheva S V 2001 Biogeochemical Processes of Methane Cycle in the Soils, Bogs, and Lakes of Western Siberia *Microbiology* **70** 175–85
[4] Glagolev M V, Sabrekov A F, Kleptsova I E, Filippov I V, Lapshina E D, Machida T and Maksyutov Sh Sh 2012 Methane emission from bogs in the subtaiga of Western Siberia: The development of standard model *Eurasian Soil Sci.* **45** 947–57
[5] Sabrekov A F, Glagolev M V, Kleptsova I E, Machida T and Maksyutov S S 2013 Methane Emission from mires of the West Siberian taiga *Eurasian Soil Sci.* **46** 1182–93
[6] Harriss R C, Sebacher D I and Day F P 1982 Methane flux in the great dismal swamp *Nature* **297** 673-74
[7] Lohila A *et al.* 2016 Large contribution of boreal upland forest soils to a catchment-scale $CH_4$ balance in a wet year *Geophys. Res. Lett.* **43** 2946–53
[8] Shoemaker J K, Keenan T F, Hollinger D Y and Richardson A D 2014 Forest ecosystem changes from annual methane source to sink depending on late summer water balance *Geophys. Res. Lett.* **41** 673–79
[9] Megonigal J P and Guenther A B 2008 Methane emissions from upland forest soils and vegetation *Tree physiol.* **28** 491–98
[10] Hutchinson G L and Mosier A R 1981 Improved soil cover method for field measurement of nitrous oxide fluxes *Soil Sci. Soc. Am. J.* **45** 311–16
[11] Terent'eva I E, Sabrekov A F, Glagolev M V, Lapshina E D, Smolentsev B A, Maksyutov Sh Sh 2017 A new map of wetlands in the southern taiga of the West Siberia for assessing the emission of methane and carbon dioxide *Water Resources* **44** 297–307
[12] Sabrekov A F, Glagolev M V, Alekseychik P K, Smolentsev B A, Terentieva I E, Krivenok L A and Maksyutov S S 2016 A process-based model of methane consumption by upland soils *Environ. Res. Lett.* **11** 075001
[13] Churkina A I, Glagolev M V, Ilyasov D V, Terentieva I E, Sabrekov A F, Maksyutov S S 2017 Variable regime of methane emission in periodically flooded forest *Carbon Balance of Western Siberian Mires in the Context of Climate Change* ed E D Lapshina and N P Mironycheva-Tokareva pp 13–17
[14] Arah J R M and Stephen K D 1998 A model of the processes leading to methane emission from peatland *Atmos. Environ.* **32** 3257–64
[15] Revut I B 1972 *Soil physics* (Leningrad: Kolos) p 203
[16] Molchanov A G 2017 Gas exchange of carbon dioxide from the surface of Sphagnum in boggy pine forests in southern taiga *Environmental dynamics and global climate change* **8** 43-54
[17] Bonan G B 1991 Atmosphere-biosphere exchange of carbon dioxide in boreal forests *J. Geophys. Res. Atmos.* **96** 7301–12
[18] Milyukova I M, Kolle O, Varlagin A V, Vygodskaya N N, Schulze E D and Lloyd J 2002 Carbon balance of a southern taiga spruce stand in European Russia *Tellus B* **54** 429–42
[19] Raich J W, Potter C S and Bhagawati D 2002 Interannual variability in global soil respiration, 1980–94 *Global Change Biol.* **8** 800–12

# Decomposition rate of peat-forming plants in the oligotrophic peatland at the first stages of destruction

**L G Nikonova*[1], E A Golovatskaya[1] and N N Terechshenko[2]**

[1]Institute of Monitoring of Climatic and Ecological Systems of the Siberian Branch of the Russian Academy of Sciences, 634055, 10/3, Academichesky Ave., Tomsk, Russia
[2]National Research Tomsk State University, 634050, 36 Lenin Ave., Tomsk, Russia

*Email: lili112358@mail.ru

**Abstract**. The research presents quantitative estimates of the decomposition rate of plant residues at the initial stages of the decay of two plant species (*Eriophorum vaginatum* and *Sphagnum fuscum*) in a peat deposit of the oligotrophic bog in the southern taiga subzone of Western Siberia. We also studied a change in the content of total carbon and nitrogen in plant residues and the activity of microflora in the initial stages of decomposition. At the initial stage of the transformation process of peat-forming plants the losses of mass of *Sph. fuscum* is 2.5 times lower then *E. vaginatum*. The most active mass losses, as well as a decrease in the total carbon content, is observed after four months of the experiment. The most active carbon removal is characteristic for *E. vaginatum*. During the decomposition of plant residues, the nitrogen content decreases, and the most intense nitrogen losses were characteristic for *Sph. fuscum*. The microorganisms assimilating organic and mineral nitrogen are more active in August, the oligotrophic and cellulolytic microorganisms – in July.

## 1. Introduction

The role of peatlands in the global carbon cycle are discussed by a number of researches. The peatlands are the unique terrestrial ecosystems, which act as a long-term sink of atmospheric carbon. Now the process of carbon accumulation dominates in native peatlands. The peatlands covering only 3% of the Earth's land area hold the equivalent of half of the carbon that is contained in the atmosphere as $CO_2$ [1, 2]. The low rate of decomposition of plant residues in peat bogs is the main mechanism allowing efficient accumulation of carbon in the peat deposits. The knowledge of quantitative indices characterizing the cycle of chemical elements in bog systems is necessary for studying the mechanisms of wetlands stability and productivity and for the substantiation of the evolution forecast for wetlands caused by the changes in the environment. The processes of transformation of plant residues in mineral soils of different natural zones have been thoroughly studied [3–8]. However, the studies devoted to the dynamics of decomposition of peat-forming plants are relatively small [4, 9–19]. The rate of decomposition of peat-forming plants depends on many factors, the main ones being the environmental conditions, the chemical composition of peat-forming plants and the activity of microorganisms [4, 19]. An important role in the study of the decomposition rate of plant residues in a peat deposit is played by the analysis of the first decomposition stages as during this period the destruction occurs most intensively.

The purpose of this work is to study the dynamics of decomposition processes at the first stages of decay of peat-forming plants in the bog ecosystems of the Southern taiga subzone of Western Siberia.

## 2. Material and methods

The researches were carried out in the territory of the Ob'-Tom' interfluve in the oligotrophic bog "Timiryazevskoe" (Tomsk district, Tomsk region) in the pine-dwarf shrub-*Sphagnum* phytocenosis. A specific point of research was pine-shrub-*Sphagnum* phytocenosis. The chosen bogs are typical for the southern taiga subzone of Western Siberia. We studied plant litter decomposition and nutrient release for two peat-forming plants: vascular plant species (*Eriophorum vaginatum*) and moss species (*Sphagnum fuscum*) using litterbag technique [4].

The litter of *E. vaginatum* and *Sphagnum* residues (10 cm from the heads of sphagnum mosses) was dried in laboratory up to air-dry state, and its portions of 5-6 g were placed into nylon bags (15×15 cm). In May 2016, the prepared samples were put into the peat soil to the depth of 10 cm from the moss surface. The samples were taken out of the peat soil each month from June to September 2016. The samples were dried in laboratory to the air-dry condition, weighed to determine their dry mass and calculate the losses.

The losses of the plant mass were calculated as a percentage of the initial sample weight according to the following formula:

$$\text{Loss (\%)} = \frac{(M_0 - M_t)}{M_0} \times 100, \qquad (1)$$

where $M_0$ is the mass of the initial sample, $M_1$ is the mass of the sample after the time $t$ (1, 2, 3 and 4 months).

In each sample the contents of carbon, nitrogen were determined before and after decomposition in accordance with [4, 20–22], and the losses of their initial content were calculated by the formula:

$$\text{Loss of an element (\%)} = \frac{(X_0 M_0 - X_t M_t)}{X_0 M_0} \times 100, \qquad (2)$$

where $X_0$ is the initial content of the element, mg/g, $X_t$ is the content of the element after time $t$ (1, 2, 3 and 4 months).

The total number of microorganisms in decomposing plant residues was studied using direct microscopy method. The activity of microorganisms was studied using Koch method. The number of microorganisms involved to the nitrogen cycle was determined on the GRM medium (for using organic nitrogen sources) and starch ammonia agar (for using organic mineral nitrogen sources). The number of microorganisms involved to transformations of carbon compounds was determined on the Hutchinson-Clayton medium.

## 3. Results

The results of the experiment showed that the rate of plant residues decomposition was rather low in the first three months perhaps because of the weather conditions in the 2016 vegetation period (very arid). The rate of decomposition increased only by the fourth month and reached 18.5% in the *Sph. fuscum* samples and 38.5% in *E. vaginatum* samples (figure 1). The *E. vaginatum* samples decomposed 2.0-3.5 times faster than the mosses samples at all stages of transformation. The maximum difference was observed in the first month of the experiment.

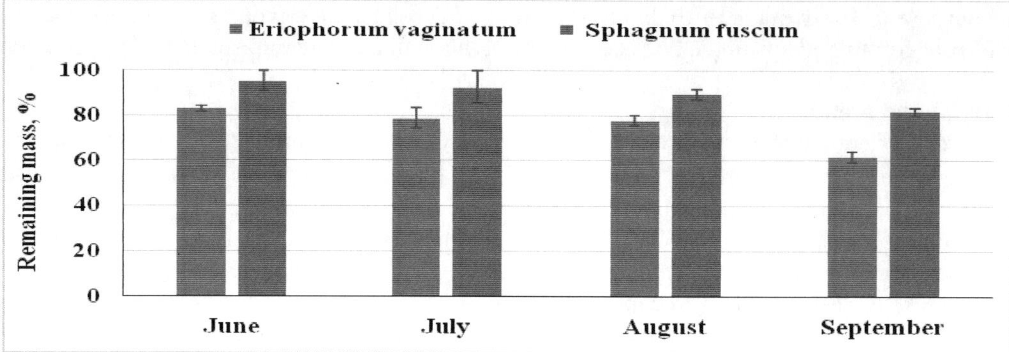


**Figure 1.** Remaining mass during decomposition of peat-forming plants.

The investigated plant residual species differ in the content of carbon and nitrogen: *E. vaginatum* is characterized by a higher content of carbon (44%) and nitrogen (0.81) compared to *Sph. fuscum* – 41% and 0.51%, respectively. During decomposition of plant samples there was a carbon and nitrogen carrying out. The maximum losses of carbon for 4 months were received to *E. vaginatum* (37%) (figure 2).

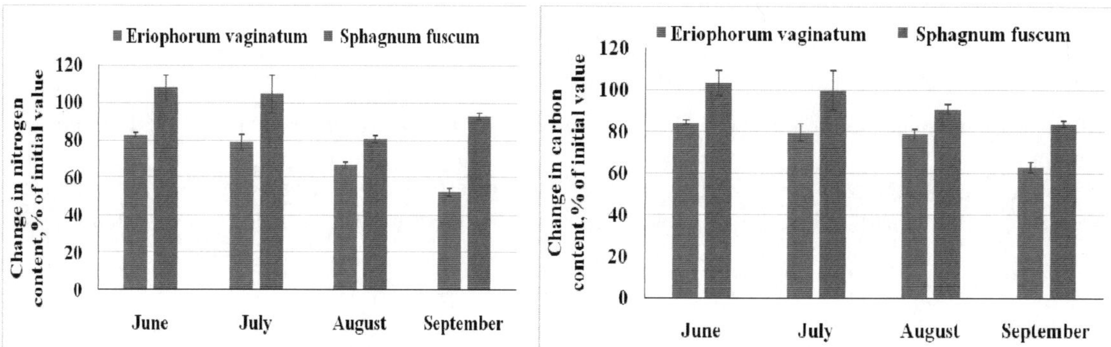


**Figure 2.** Change in carbon and nitrogen content in plant samples.

Fast removal of carbon in *E. vaginatum* samples occurred in the first and fourth months (16%), during the second and the third months the removal of carbon was very slow (1-5%).In the first two months of the experiment the carbon loss in *Sph. fuscum* samples was not observed, but in the end of the experiment the removal of carbon reached 16% of the original content (figure 2). The maximum number of cellulolytic microflora in *Sph. fuscum* residues was observed only in the beginning of the growing season – in June, but in September the activity of cellulolytic microflora increased again (table 1).

The dynamics of the nitrogen content in the studied species under their decomposition were different. During the decomposition of plant residues, both the mineralization of nitrogen and its immobilization (accumulation) may occur. According to [23], the process of N accumulation may take place at the first stages of decomposition due to the destruction of nitrogen-free compounds, nitrogen fixation and nitrogen consumption by fungal hyphae from the soil. Nitrogen removal starts only after reaching a critical concentration of nitrogen in the plant residues.

In the first month nitrogen losses of about 17% of the original content were observed in the process of *E. vaginatum* samples decomposition. For *Sph. fuscum*, a direct relationship was found between the amount of ammonifiers and the decrease in the nitrogen content. The maximum number of these microorganisms in the samples of *Sph. fuscum* was observed in August. In September, the number of microorganisms absorbing both mineral and organic forms of nitrogen decreases in comparison with August and that explains the relative increase in N content in *Sph. fuscum* samples (table 1).

IOP Conf. Series: Earth and Environmental Science **138** (2018) 012013 doi:10.1088/1755-1315/138/1/012013

For plant residues *Sph. fuscum* the total number of microorganisms significantly increased by August – September. For *E. vaginatum* samples and peat samples taken from the plant locations (10 cm), the maximum activity of microorganisms was observed in July. Probably the main reason for increase in the total abundance of microorganisms by the end of summer was decrease in the level of bog water and improvement of peat aeration as a result of very arid weather conditions from June to July in 2016. An improvement in the aeration conditions of the peat deposit in August is also indicated by an increase in the number of microorganisms that assimilate both organic and mineral nitrogen. The increase in the number of microorganisms occurred mainly due to fungi, which was very demanding for the presence of sufficient oxygen in the substrate (table 1).

As it is well known, oligotrophic microorganisms may use nitrogen or carbon nutrition from substrate containing very low concentrations of these elements [24]. Among the studied plant samples the largest number of oligotrophic microorganisms was found in *E. vaginatum* samples. The maximum number of these microorganisms was fixed in July. For the *Sph. fuscum* samples the maximum number of oligotrophic microorganisms was observed in September (table 1). It is noteworthy that the dominant group among the oligotrophic microorganisms was bacteria, the fungi share was much less.

**Table 1.** Total number of microorganisms in peat and plant residues.

| Sample | Month | Microorganisms assimilating organic nitrogen (GRM agar) (N x $10^4$ CFU* / 1 g of d.s.** | | | Microorganisms assimilating mineral nitrogen (SAA) (N x $10^4$ CFU* / 1 g of d.s.** | | | Oligotrophic microorganisms (hungry agar) N x $10^4$ CFU* / 1 g of d.s.** | Cellulosolytic microflora, (Hutchinson-Clayton medium) N x $10^4$ CFU* / 1 g of d.s.** | |
|---|---|---|---|---|---|---|---|---|---|---|
| | | Total number | Bacteria | Fungi | Total number | Bacteria | Fungi | Total number | Total number | Myxobacteria |
| Peat*** | 1 | 1665 | 1643 | 22 | 1264 | 1234 | 301 | 216.8 | 0.261 | <10 |
| | 2 | 27090 | 24147 | $<10^3$ | — | — | — | 23174.0 | 1.206 | <10 |
| | 3 | 9951 | 2552 | 7399 | 21926 | 21378 | 548 | 3522.7 | 0.052 | <10 |
| | 4 | 7715 | 7714 | 1 | 10322 | 10250 | 72 | 5280.1 | <10 | <10 |
| *Sph. fuscum* | 1 | 1272 | 1262 | 10 | 195 | 195 | $<10^3$ | 1098.78 | 0.229 | 0.115 |
| | 2 | 7980 | 7980 | $<10^3$ | — | — | — | 12873.0 | 0.007 | <10 |
| | 3 | 23341 | 11646 | 11695 | 23811 | 22526 | 1285 | 2679.5 | 0.009 | <10 |
| | 4 | 10236 | 10065 | 171 | 256 | 256 | $<10^3$ | 37361.4 | 0.043 | <10 |
| *E. vaginatum* | 1 | 2195 | 2146 | 49 | 3733 | 3726 | $<10^3$ | 1178.8 | 0.42 | <10 |
| | 2 | 28140 | 71 | 28070 | — | — | — | 77571.0 | 0.63 | 0.51 |
| | 3 | 16593 | 658 | 15933 | 23311 | 23016 | 296 | 3173.4 | 0.157 | <10 |
| | 4 | — | — | — | — | — | — | — | — | — |

Footnote: CFU* – colonies forming units; d.s.** – dry soil; *** – Sphagnum residues (10 cm from the moss surface); "—" – no data

## 4. Conclusion

The decomposition of *Sph. fuscum* is a major contributor to the peat formation process at the initial stages. Decreasing of the total carbon content, as well as mass loss, was most active in the third and fourth months of the experiment. *E. vaginatum* residues were characterized by the most intensive carbon removal. During the decomposition of plant residues the most intensive nitrogen losses were characteristic for *E. vaginatum*. The nitrogen content in *Sph. fuscum* residues was increasing during two months but then its content decreased. The microorganisms assimilating organic nitrogen were most active in the peat samples and in the *Sph. fuscum* remains. The *E. vaginatum* samples are characterized by the prevalence of microorganisms assimilating inorganic nitrogen. The microorganisms assimilating organic and mineral nitrogen were most active in August and the oligotrophic and cellulosolytic microorganisms were more active in July.

**References**

[1] Vompersky S E 1994 The role of marshes in the carbon cycle *Biogeocenotic features of bogs and their rational use* (Moscow: Nauka) pp 5–37
[2] Succow M and Joosten H 2001 *Landschafts ökologische Moorkunde* (Stuttgart)
[3] Aristovskaya T V 1965 *Microbiology of Podzolic Soils* (Moscow: Nauka)
[4] Kozlovskaya L S, Medvedeva V M and Piavchenko N I 1978 *Dynamics of organic matter in the process of peat formation* (Leningrad: Nauka) p 176
[5] Semenov V M, Ivannikova L A, Kuznetsova T V, Semenova N A and Khodzhaeva A K 2007 *Eurasian Soil Sci.* **40** 1208–1216
[6] Stenina T A 1964 *Pochvovedenie* **1** pp 95–102
[7] Tepper E Z, Komarevtseva L T and Simakova N S *Izv. Timiryazev S-kh. Akad.* **1** pp 121–126
[8] Ushakova G I 2000 *Eurasian Soil Sci.* **33** 881–886
[9] Bambalov N N, Khoruzhik A V, Lukoshko E S and Strigutskii V P 1990 Transformation of dead plants in wetland biogeocenoses *Role of the Experiment and Mathematical Modeling in Analysis of Forest and Wetland Biogeocenoses* (Moscow: Nauka) pp 53–63
[10] Golovatskaya E A and Nikonova L G 2013 *Vestn. Tomsk. Gos. Univ., Biol.* **3 (23)** 137–151
[11] Golovatskaya E A and Nikonova L G 2017 *Eurasian Soil Science* **50** (5) 580–588
[12] Parshina E K 2009 *Destruction of plant litter in peatland ecosystems of taiga and forest-tundra zones of Western Siberia* Ph.D. Dissertation (Tomsk)
[13] Bragazza L, Siffi C, Iacumin P and Gerdol R 2007 *Soil Biol. Biochem.* **39** 257–267
[14] Hajek T 2009 *Boreal environmental research* **14** 947–958
[15] Jiroušek M 2012 *Ecology of bogs in the Sudetes Mts.* Ph.D. Dissertation (Brno) 124 pp
[16] Peltoniemi K, Strakova P, Fritze H, Iraizoz P A, Pennanen T and Laiho R 2012 *Soil biology and biochemistry* **51** 20–34
[17] Turetsky M R, Crow S E, Evans R J, Vitt D H and Wieder R K 2008 *Journal of Ecology* **96** 1297–1305
[18] Verhoeven J T A and Arts H H M 1992 *Aquat. Bot* **43** 365–377
[19] Waddington J M, Rochefort L and Campeau S *Wetlands ecology and management* **11** pp 95–95
[20] Chernov I Yu (Ed) 2013 *Activity of Microbial Complexes in Upper Peatbogs: Analysis of Slow Destruction of Peat* (Moscow: KMK) 128 pp
[21] Sokolov A V 1975 *Agrochemical methods of soil investigation* (Moscow: Nauka) 656 pp
[22] Ponomareva V V and Plotnikova T A 1975 *Methodological Recommendations for Determination of the Composition of Humus in Peat and Mineral Soils* (Leningrad: Nauka) 105 pp
[23] Bazilevich N I and Titlyanova A A 2008 *Biotic circulation on five continents: nitrogen and ash eltmtnts in natural terrectrial ecosystems* (Novosibirsk) 381 pp
[24] Mishustin E N 1975 *Associations of soil microorganisms* (Moscow: Nauka) 107 pp

IOP Conf. Series: Earth and Environmental Science **138** (2018) 012014 doi:10.1088/1755-1315/138/1/012014

# Peatland development and paleoclimate records from the Holocene peat archive in the foothills of the Eastern Sayan Mountains

**A B Rodionova, A V Grenaderova**

Department of Ecology and Environmental Studies, Institute of Ecology and Geography, Siberian Federal University, 79 Svobodny pr., 660041 Krasnoyarsk, Russia

Email: rodionovaab@yandex.ru, grenaderova-anna@mail.ru

**Abstract**. Plant macrofossils data were used to identify the successive peatland communities during the last 3500 years in the floodplain of the Mana River (foothills of the Eastern Sayan Mountains). The reconstruction of the peatland development indicated that the peatland in the Mana River basin formed about 3500 years ago. The peatland formed as a result of overgrowing floodplain and water logging of terrace lows. The authors observed three successive changes: birch forest with sedge and hypnum mosses in the second half of the Subboreal period, wood-marsh plant association at the start of the Middle Subatlantic period (1600 years BP), the herb-wort phytocoenosis with inclusions of mezoeutrophycal plant species have been growing since the Late Subboreal period (950 years BP).

## 1. Introduction

Peatlands are an important natural archive for past climatic changes, primarily due to their sensitivity to changes in the water balance and the dating possibilities of peat sediments. The climatic changes throughout the Holocene have been reconstructed from peat, using a wide array of biological and other proxies. Many different proxy-indicators may be derived from peat cores [2].

Peat-based climatic and environmental reconstructions are currently available from many sites in Yenisei Siberia, mainly for its northern territories. The purpose of this research is to study some features of peatland development and environmental reconstructions from the Holocene period in the south part of Yenisei Siberia (foothills of the Eastern Sayan Mountains).

The study area was the peatland "Narva" in the floodplain of the Mana River. The main method of the research was macrofossils analysis of peat. This analysis can be used to reconstruct the development of the local vegetation and surface wetness on peatlands, and thus to elucidate successional processes. To interpret these proxies by macrofossil analysis, the current botanical composition of the bog and the ecological behavior of different plant species were used. The analysis of plant macrofossils in the peat, based on the study of the flora and vegetation in a particular place over a period of time, has allowed us to reconstruct the environmental changes that have occurred since the Late Glacial Period. The macrofossil types were identified with Kac and Kac Atlas [8]. The reconstruction of the bog surface wetness was performed on the basis of the analysis of plant macrofossils. After that, we used the ecological scales of moisture and reconstructed surface wetness for the entire period of the peatland formation.

IOP Conf. Series: Earth and Environmental Science **138** (2018) 012014 doi:10.1088/1755-1315/138/1/012014

## 2. Materials and methods

The Narva peatland is located at the foothills of the Eastern Sayan Mountains, Yenisei Siberia, Russia, at a distance of 20 km of the southern boundary of Kansk forest-steppe. The peatland has a width of 2 km, a length of about 6 km, and is extended along the right bank of the Mana River (Yenisei's inflow). The ground water level has been recorded at around 20 cm. The surface vegetation mostly comprises eutrophic (bog) assemblages. The present vegetation is a mixed herb-moss forest. The shrub layer includes *Oxycoccus microcarpus* Turcz. The herbal layer consist of sedges (such as rhizomatous species of sedge (*Carex cespitosa* L., *C. appropinquata* L.) and *Equisetum palustre* L. The ground layer is largely dominated by the Hypnales (*Tomentypnum nitens* Hedw., *Aulacomnium palustre* Hedw.) and Sphagnales (*Sphagnum warnstorfii* Russ).

A cross-section was laid in the peatland for the watershed part toward the shore line of the Mana River. The cross-section length was 550 m. The drilling sites were chosen in places with different phytocenoses. We carried out drilling of peat cores in three locations. The altitude at point 1 was 380 meters. The difference in altitude from point 1 to the riverbed was 1 m.

Three peat cores (Narva 1, Narva 2, Narva 3) were collected in July 2011 from lawn microforms with a Box core. The collected cores were wrapped in plastic gutters until required for analysis. The types of sediments were identified in the laboratory.

The maximum thickness of peat deposits (1.50 m) was recorded for a hillside (20 meters apart from the base of the Mana River first terrace). The peat deposit consisted of eutrophical peat types: wort, sedge-wort, wood-grass and horsetail.

The peat samples (5 $cm^3$) from cores Narva 1, Narva 2, Narva 3 were taken at 5 cm increments. They were heated in 5% KOH solution and rinsed with distilled water through a 0.5 cm sieve. The analysis was carried out by counting macrofossil pieces, using available identification keys [8, 14] and by estimating peat components as volume percentages of the total sample. The volume percentages of different remained vegetation and Sphagnum section were estimated in step of 5%. The identification of Sphagnum to species level was performed separately based on stem leaves using specialized keys [7].

The multi-proxy research of detailed macrofossil analysis and peat stratigraphical studies provided temporal precision and increased understanding of climate changes on local and regional scales [3].

Feurdean et al. [5] suggested using combined macrofossil and pollen analysis, which assisted in understanding of the differences between local and regional flora. The lack of macrofossils from leafed trees, which were plenty in the surrounding vegetation according to the pollen data, indicated that these trees were not growing on the peatland or around its banks.

After macrofossils analysis we used the ecological scales of moisture by Ramensky and Ellenberg [4, 12] and reconstructed the surface wetness for the entire period of the bog formation based on the plant degree of moisture-loving. The ecological scale of moisture range by Ramensky has 120 points. The Ellenberg ecological scale has 12 points.

The ecological scales were about the methods of assessing a plant habitat and determining a plant community position along the gradients of ecological factors.

Ramensky placed each species on a 120-point ordinal scale according to its distribution with respect to moisture. The characteristics of the moisture regime (e.g. groundwater level, soil moisture content, and soil moisture deficit) were used in determining these classes. Ramensky's indicator values were used to estimate the value of any of these environmental factors at a particular site by averaging the indicator values for this factor of all species present. Plants often reflect temporally integrated environmental conditions and they were therefore particularly useful indicators when values averaged over time were needed. To identify the value of the environmental factor in the past the only possible approach may be to relate it to historical vegetation data. The Ramensky scale offered 5 gradations (table 1).

**Table 1** Gradations which accounted for projective cover.

| **m (abundant)** | **more than 8%** |
|---|---|
| c (rich) | 2.5-8% |
| n (moderate) | 0.3-2.5% |
| p (low) | 0.1-0.2% |
| a (sporadic) | <0.1% |

The European scale of moisture was not comparable to Ramensky scale, due to the diversity of natural areas and the range of geographical regions. The Ramensky scale covers tundra, forest- tundra, taiga, coniferous and broad-leaved forests, forest-steppe, steppe, desert. The entire 120-point scale was divided into 12 stages.

Each stage included a total of 6 to 13 points, only the first (the desert) was 17 points. The other 5 stages (starting with 77 points) were specific for intrazonal habitats and communities (peatlands, water and riparian vegetation).

**Table 2** Ellenberg scale & Ramensky scale points.

| **Ellenberg scale** | **Ramensky scale** |
|---|---|
| 1 on extremely dry soils, e.g. bare rocks | 1-17 desert |
| 2 in-between 1 and 3 | 18-30 semidesert |
| 3 on dry soils | 31-39 dry steppe |
| 4 in-between 3 and 5 | 40-46 middle steppe |
| 5 on fresh soils, i.e. under intermediate conditions | 47-52 wet steppe |
| 6 in-between 5 and 7 | 53-63 dry and fresh grassland and forest |
| 7 on moist soils which do not dry out | 64-76 wet grassland |
| 8 in-between 7 and 9 | 77-88 moist grassland |
| 9 on wet, often not well aerated soils | 89-93 marsh-grassland |
| 10 on frequently inundated soils | 94-103 peatland |
| 11 water plant with leaves mostly in contact with the open air | 104-109 sudd |
| 12, mostly totally immersed in water | 110-120 underwater plant |

We assessed humidity conditions during the whole period of peat accumulation and marked different phases in the development of peatland, and described their vegetation. Radiocarbon dating was carried out at Sobolev Institute of Geology and Mineralogy, Russian Academy of Sciences, Novosibirsk. Dates had been calibrated using the probability method and the bidecadal data set in the computer program CALIB version 7.1.

## 3. Results and discussion

Stratigraphy, radiocarbon dating and plant macrofossils were used to reconstruct the development and ecology of a small raised peatland because a peatland with a small area was most sensitive to climate change [9]. The Narva peatland area was 1020 ha, which was relatively small.

The peat layers were underlayed by sediments of the floodplain facies – sandy loam, which dated back to the middle Subboreal time (radiocarbon age was 3405 ± 167 cal. years BP in the interval 1.75-1.85 m). The floodplain facies consisted of oxbow loams which included macrofossils of sedge fabrics, horsetail, bark of spruce and birch, isolated green moss macrofossils.The birch forest with sedge, spruce and hypnum moss was developed in a high floodplain of the Mana River in the second half of the Subboreal period. Shallow bodies of water were located along the river and at some distance from it and that resulted in the general decrease in the water content and started its overgrowing. The process of peat accumulation was intensified (figure 2) in the warm conditions of the late Subboreal period. The increased degree of decomposition and low peat ash gave evidence about the overall decline of moisture conditions and rising of the river water level. These conditions led to stable conditions of the hydrological regime.

At the beginning of Subatlantic period (figure 2) *Picea obovata* Ledeb dominated in the tree layer of paleo plant formation with *Equisetum palustre* and then *Carex lasiosarpa* L. dominating in the herbal layer. The spruce forest with sedge and hypnum moss (*Tomentypnum nitens* and *Aulacomnium palustre*) grew under these conditions.

The wood-marsh plant formations were changed through herb-wort in 1600 BP. The herbal layer consisted of *Equisetum palustre*, *Calamagrostis lanceolata* L., *Eriophorum vaginatum* L., *Comarum palustre* L., *Menyanthes trifoliata* L., and sedges: *Carex cespitosa*, *C. lasiosarpa* , *C. globularis* L., *C. limosa* L.

Shrubs with *Oxycoccus microcarpus* were recorded for this period. The quantity of green moss had reached the maximum value in the content of paleo plant formation. The reduction of peat decomposition and the increase of ash content were probably due to an increase in the duration and magnitude of floods. This time could be described as cool and wet, similar to the stage Fernau (the Alps) or Aktru (the Altais) [1].

Mixed herb plant formation was pervasive in the Narva peatland from the first half of the late Subatlantic period. However, the grass layer was reduced and participation of horsetail mezoeutrophical species was increased.

The majority of peatlands had a relatively young age in Kansk forest-steppe zone (from the end of the Atlantic to the beginning of subboreal period). The process of peat accumulation dated middle Subboreal time. More ancient peatlands were only preserved far from the major rivers, in places which were protected from erosion scours in the Boreal and Atlantic periods [15].

The rate of peat accumulation was 0.55 mm / year for the Narva peatland, which was slightly above the average for Kansk forest-steppe (0.44 mm / year) and the Krasnoyarsk forest-steppe (0.31 mm / year). The maximum rate of peat accumulation (0.72) was recorded at the bottom of the peat deposit, which was formed in the Late Subboreal and Early Subatlantic periods. According to the results of a peatland study dated with a smaller step, the maximum rate was observed from the second half of the Subboreal to the beginning of Subatlantic period (1.2 mm / year in the Terteg peatland (Kansk forest-steppe), and 1.6 mm / year in Kacha peatland ( Krasnoyarsk forest-steppe) [6, 13].

Our study presented paleohydrological reconstructions from a eutrophycal peatland. Eutrophycal and ombrotrophic peatlands received their moisture inputs solely from precipitation. Such peatlands tend to be used as a source of the Holocene climate data. Long-term variations in peatland surface wetness were interpreted as changes in the regional climate. Magnan et al [10] concluded that the main paleohydrological trends from mid-to-late Holocene climate changes provided response of peatland ecosystems to climate variability.

In the end of Subboreal period we identified the transformation of wood-mash plant communities to herbal communities associated with changes in palaehidrological regime of the study area. Mithell et al.[9] concluded that the forest clearance resulted in increased evapotranspiration causing lowering of the water table on the bog and modification of the vegetation cover. This hypothesis had implications for the management of similar small raised bogs in karst-dominated landscapes (Finland and northern part of Yenisei Siberia). According to our data, the forest clearing did not occur in the study area.

The plant macrofossil and ash-content analysis showed changes in the species composition. McCaroll et al. [11] suggested that it could be difficult to differentiate between natural and anthropogenic patterns vegetation changes during the late Holocene. According to the results of our study, a high content of mineral elements in the peat cores from the last 300 years may be explained by a wide spread of agriculture in the study area.In addition, pollen analysis will be obtained in the next period of our research project. That may show changes in the regional vegetation and assist in making a conclusion about the human activity in the study area.

## 4. Conclusion

The peatland formation started in the Early Subboreal period in the Mana River valley. All stages of the peatland formation were eutrophycal. The maximum value of the moisture conditions was 98 points. This result differed from the forest-steppe peatlands where the maximums peat moisture conditions were 92-94 points. The birch forest with sedge and hypnum mosses in its herbal layer grew in the second half of the Subboreal period. An active process of overgrowing and peat accumulation started in the shallow waters along the Mana River was due to the general decrease of the water level in the river. These processes developed under warm conditions of the Late Subboreal period.

The mixed herbal plant formation replaced the wood-marsh plant association at the start of the Middle Subatlantic period (1600 years BP). The reduction in the degree of peat decomposition and the increase in the ash content indicated a longer and large-scale floods. This stage could be described as cool and moist.The herb-wort phytocoenosis with inclusions of mezoeutrophycal plant species have been growing since the Late Subboreal period (950 years BP) with the high moisture conditions (peatland moisture condition) of the peatland.

## Acknowledgements

The reported study was funded by RFBR according to the research project № 16-04-00796 A

## References

[1] Bezrukova E V, Vershinin K E, Letunova P P, Orlova L A, Krapivina S M, Chepinoga V V, Verhozina A V, Dudareva N V and Abzaeva A A 2004 *Botanicheskij zhurnal* **89** pp 221–232
[2] De Jong R, Blaauw M, Chambers F M and Christensen T R 2010 *Climate and peatlands* (Netherlands: Springer) pp 85–121
[3] Edvardsson J, Linderson H, Rundgren M and Hammarlund D 2012 *Journal of Quaternary Science* **27 (6)** pp 553–563
[4] Ellenberg H 1988 *Vegetation ecology of central Europe* (Cambridge: Cambridge University Press) pp 15–24
[5] Feurdean A N and Bennike O 2008 *Studia UBB Geologia* **53 (1)** pp 5–10
[6] Grenaderova A V 2005 *Dinamika bolot Krasnoyarskoj i Minusinskoj lesostepej: avtoref. diss. kand. geogr. nauk* (Barnaul) p 22
[7] Ignatov M S and Ignatova E A 2003 *Moss flora of the Middle European Russia* (Moscow: KMK Scientific Press Ltd.) pp 609–960
[8] Kac N Ya, Kac S V and Skobeeva E I 1977 Atlas rastitelnyh ostatkov v torfah (Moscow: Nedra) pp 372
[9] Mitchell E A, van Der Knaap W O, van Leeuwen J F, Buttler A, Warner B G and Gobat J M 2001 *The Holocene* **11(1)** pp 65–80
[10] Magnan G and Garneau M 2014 *Journal of Quaternary Science* **29(3)** pp 209–220
[11] McCarroll J, Chambers F M, Webb J C and Thom T 2017 *Quaternary International* **432 (A)** pp 39–47
[12] Ramenskij L G 1956 *Ekologicheskaya ocenka kormovyh ugodij po rastitel'nomu pokrovu* (Moscow: Gosudarstvennoe izdatelstvo selskohozyajstvennoj literatury) pp 57–67
[13] Rodionova A B, Grenaderova A V 2016 *Vestnik Krasnoyarskogo gosudarstvennogo agrarnogo universiteta* **4** pp 65–72
[14] Tyuremnov S N 1976 *Torfyanye mestorozhdeniya* (Moscow: Nedra) p 464
[15] Yamskih A F 1993 *Osadkonakoplenie i terrasoobrazovanie v rechnyh dolinah Yuzhnoj Sibiri* (Krasnoyarsk: KGPI) pp 117–128
[16] Kutenkov S A 2013 *Proceedings of the Karelian Scientific Center of the Russian Academy of Sciences* **6** pp 171–176

IOP Conf. Series: Earth and Environmental Science **138** (2018) 012014 doi:10.1088/1755-1315/138/1/012014

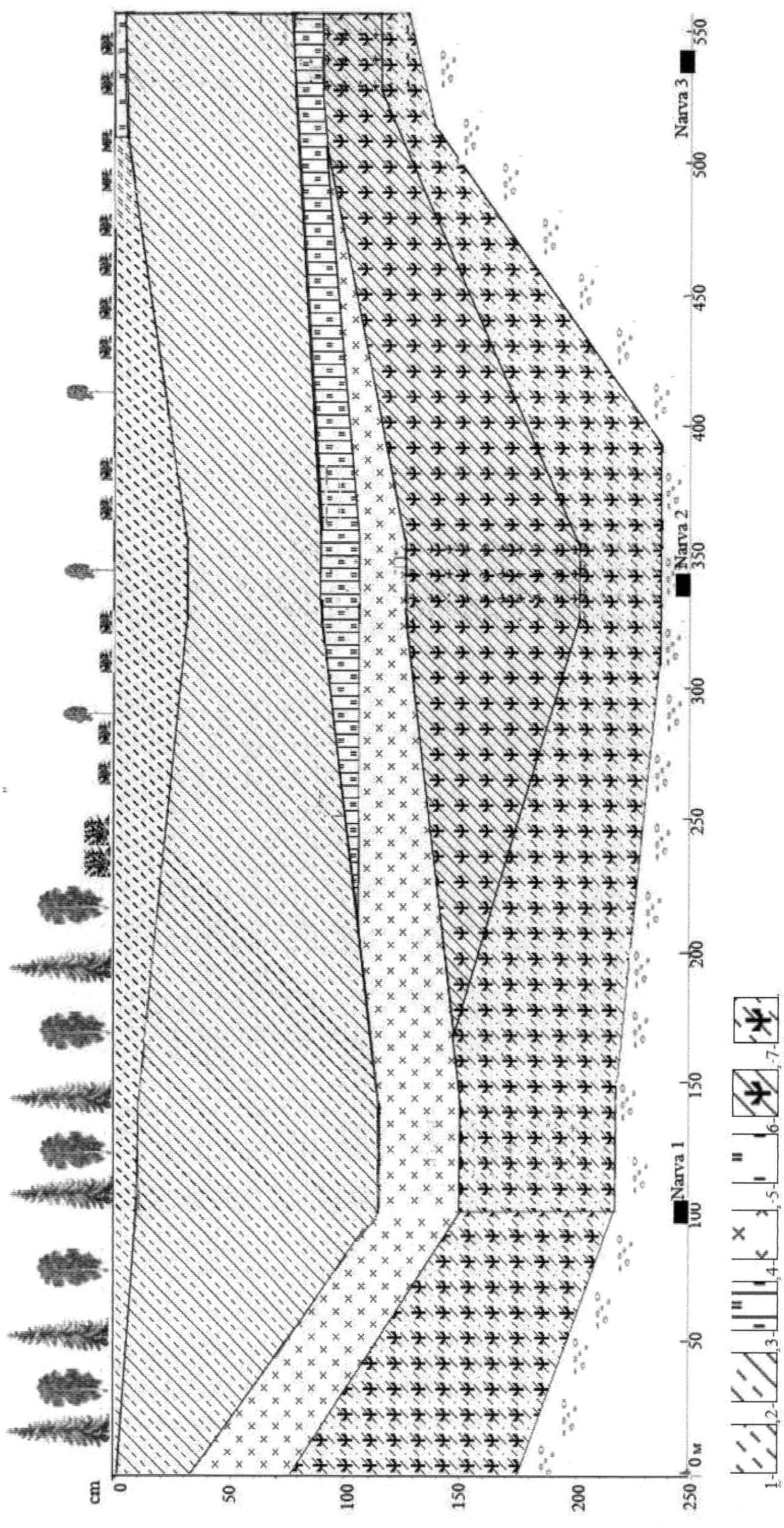


**Figure 1**. Schematic structure of the peatland deposits "Narva". Peat types: 1 – hypnum, 2 – sedge-hypnum, 3 – wood-herbal, 4 – equisetum, 5 – herbal, 6 – clay, 7 – sand clay.

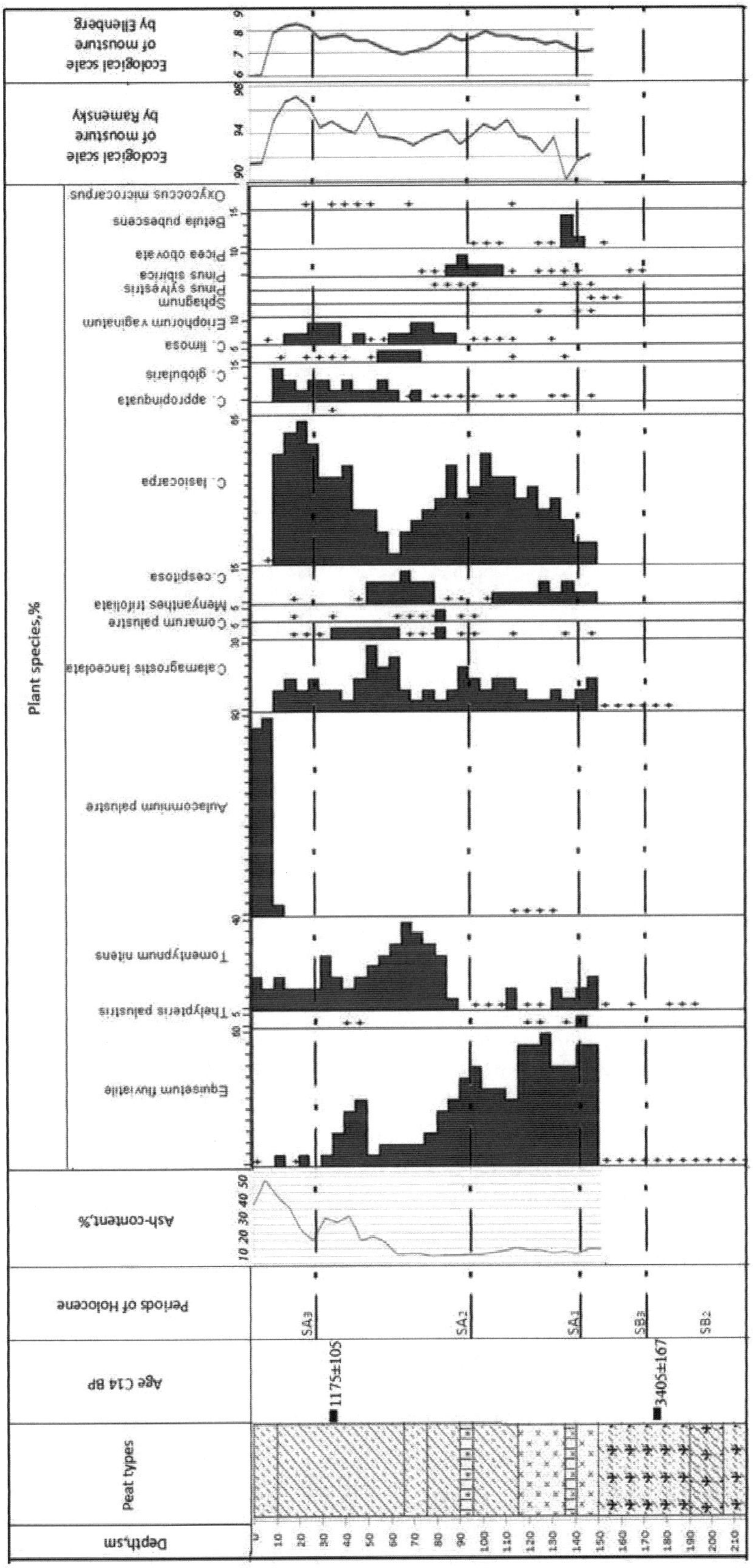


·1-, ·2-, ·3-, ·4-, ·5-, ·6-, ·7-.

**Figure 2**. Peat types, ash content, plant species in peat samples and ecological scale of moisture. Peat types: 1 – hypnum, 2 – sedge-hypnum, 3 – wood-herbal, 4 – equisetum, 5 – wood-equisetum, 6 – clay , 7 – sand clay.

IOP Conf. Series: Earth and Environmental Science **138** (2018) 012015 doi:10.1088/1755-1315/138/1/012015

# Objectives and conditions for protection of unique peatlands in the south of Western Siberia in connection with issues of conservation and use of unique reed mire in suburbs of Tomsk

**N M Semenova**

Tomsk State University, 36, Lenin Avenue, Tomsk, 634050, Russia

Email: nmsemnv@mail.tomsknet.ru

**Abstract**. The article covers the issues of protection of the wetlands located in different natural and economic conditions of Western Siberia. The main focus is on preservation and wise use of the peatlands in the southern part of the West Siberian Plain, which was greatly altered and transformed during the long-term economic development. The field research was conducted in the southern suburbs of Tomsk. A unique peatland, which is an object of a special protection effort, is described in this work. The peatland develops in the conditions of a rich groundwater feed. The peatland deposits are a mixture of peat and travertine. The extension area and the maximum thickness of organo-mineral deposits were determined. The composition of the peat deposits waters and feeding springs was studied. The specific properties of the surface waters formed under the influence of the wedging-out underground waters of the Paleozoic horizons were considered. The flora and vegetation of the study area were analyzed. A list of rare and endangered plant species was compiled. The ecological and social value of the investigated territory was substantiated. As a result we defined pertinent measures needed to protect the natural landscape and to use wisely its resources.

## 1. Introduction

The Western Siberia is widely known for its giant watershed peatlands that attained international fame as gigantic filters-absorbers of atmospheric pollutants, water and carbon storage of global significance. Here we talk about the peatlands of the northern half of the region, where they utterly dominate in the structure of surrounding landscape directing and transforming the order of nature development [1, 2]. However, the peatlands of the southern part of the Western Siberia, despite their less massive extension and relatively small individual sizes, present some unique features of equal significance [3]. Initially representing the economic interest, mainly as a source of valuable organic or carbonate raw materials, they gradually began to be perceived as the resources of a special environmental purpose.

The tasks of protecting the unique peatlands in the south of the Western Siberia usually correlate with the tasks of the biota conservation, including the rare and endangered species of animals and plants, which are permanently, temporarily or forcedly associated with these habitats [4]. The structure and features of the peatland sediments present a particular nature protection interest. The third component of the protection motivations is formed in a specific natural and economic environment in connection with certain forms of use of peatland landscapes and their immediate surroundings [5]. The same circumstances determine the possibility and conditions for protection of individual most important peatlands.

Thus, the purpose of this article is to analyze the existing prerequisites and conditions for the establishing of special protection of a small reed mire on the outskirts of Tomsk.

IOP Conf. Series: Earth and Environmental Science **138** (2018) 012015 doi:10.1088/1755-1315/138/1/012015

The following 3 tasks were set to achieve the objective:

1. Study the vegetation of the research area and define rare and endangered plant species and specifics of their localization related to a type of habitat.
2. Define the specifics of the studied reed mire in connection with the geologic and geomorphologic and hydrogeologic conditions of its development.
3. Establish a possibility of special protection of the mire in connection with the common practice of use of the natural resources and landscape in this area.

## 2. Study Area

The territory under study is administratively located within the boundaries of the Tomsk oblast in the suburban area of the city of Tomsk. Geomorphologically, it is an area on the right bank of the river Tom, which includes the lower reaches fragment of the Basandayka river valley.

Basandayka is a small right-bank tributary of the river Tom, which limits from the south the main residential development around the city of Tomsk. The distance from the residential areas of Tomsk along the highway is about 7 km (figure 1), from the river Tom in a straight line is about 4 km.

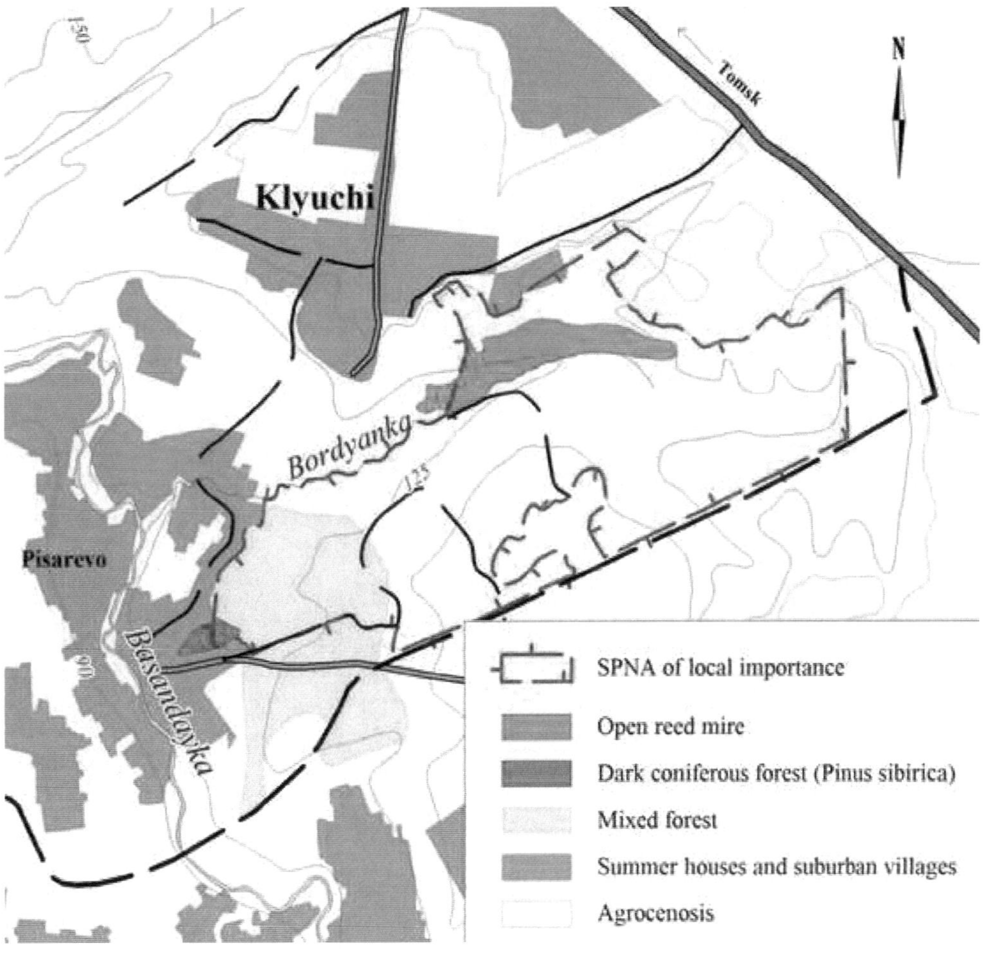


**Figure 1.** Study area [5].

The terrain is strongly dissected and very contrasting. The elevations vary from 90 m to 190 m. The landscapes are very picturesque. A distinguishing characteristic of the territory is the abundance of springs. The area has long been popular in a recreational sense. The health rehabilitation centre

'Klyuchi' located there is the oldest health resort in Tomsk and at the same time a modern well-known medical and health organization in Siberia.

The natural vegetation was transformed by human activities, except for the ravines and valleys of small tributaries of the river Basandayka, the bottoms of which are generally swamped. An extended valley bottom of one of the tributaries of Basandayka – the river Bordyanka is occupied by a reed fen of a rich groundwater feed (figure 1).

The great portion of the fen is a part of a specially protected natural area of local significance, which was established there in 2010 at the request of the local residents objecting to the uninterrupted development of recreationally valuable land. The protected area is about 100 hectares in size and is not effective for solving pressing research and applied tasks of the environmental and landscape protection in this area.

## 3. Materials and methods

The data describing the conditions and specifics of the studied area were obtained by analysing the archived materials and through field research. Among the archives we studied as well the geological archives of the Tomsk oblast. The field studies were conducted from 2009 to 2016 and included landscape and geobotanical research, landscape mapping, defining rare and endangered plant species, soil probing and natural water sampling for an analysis in the accredited laboratories of Tomsk.

The geobotanical research was conducted using the route method. The herbarium samples were identified in the Krylov Herbarium of the Tomsk State University. The stratigraphy specifics and total reserves of the peat and travertine deposit of the studied mire were described using preliminary data from geological scouting activities. The data on extension of travertines were revised during the field studies. The map of the geological structure of this area was compiled using the software MapInfo and a 1:25 000 scale topographic map.

The content of the chemical elements in the water samples analyzed in this article in connection with the peculiarities of local geochemical conditions was determined in the laboratory of the 'Analytical Centre of Geochemistry of Natural Systems' of the Tomsk State University using the ICP-MS (Agelent 7500 cx, research associates Rabtsevich E., Nikitina E.) in accordance with HCAM-XM No. 480-X [6].

The development trends and prospects of nature management in the studied area were assessed using (1) the available data on the reserves and quality of the most valued resources, (2) the evaluation of current or possible ways of using them, (3) the study of the outcomes from the previous research of natural waters conducted by the Tomsk Research Institute of health resorts and physiotherapy

## 4. Results and discussion

The results of this work correspond to the main tasks of peatland protection in the south of the Western Siberia. They confirm the natural value of the reed fen in the valley of the river Bordyanka near the Klyuchi village in the southern suburbs of Tomsk and the need for its special protection along with the neighbouring territories. The study shows the peculiarities of the nature management in the area and the forms and methods of use of natural resources that emphasize the unique traits of the local landscape.

### 4.1. *Biological and landscape diversity*

The valley of the river Bordyanka is a unique natural habitat in a picturesque summer-house nearby suburbs of Tomsk with a symbolic name 'Klyuchi'. The local landscape is dominated by the slope surfaces with numerous springs and diverse vegetation: meadow-steppe, forest-meadow, taiga, swampy. About 250 species of higher vascular plants, 23 of which are rare and endangered species, are registered there along a relatively small the specially protected area 'The Bordyanka river valley' (table 1). In terms of habitat diversity, floristic richness and repletion with rare species the site is considered unique in botanico-geographical respect not only for Tomsk, but for the oblast on the whole [5]. Almost in the city boundaries one can encounter there, yet in high abundance, *Cyprepedium macranthon* Sw. and *Cyprepedium guttatum* Sw. Actually, the reed mire communities are the most rare in the nature conservation cluster of the Tomsk oblast [7].

The reed (*Phragmites australis* (Gav.) Tril) on the mire of the river Bordyanka valley forms dense thickets and a heavy bed from its dead stems and is the dominant species and displaces other plants. Under the canopy of less closed reed communities there can be found horsetails, marsh willowherb (*Epilobium* palustre L.) and yellow loosestrife (*Lysimachia vulgaris* L.). In waterlogged micro-depressions dominant are sedge (*Carex acuta* L.) and moisture-loving species: marsh marigold (*Caltha palustris* L.), tufted loosestrife *Naumburgia thyrsiflora* (L.) Reichenb.), marsh bedstraw, veronica, stitchwort, etc. Some areas are dominated by water horsetail (*Equisetum fluviatile* L.).

At spring, across the debris cones tussocks (*Carex cespitosa* L.) exits to the surface among the sedge, hogweed (*Heracleum dissectum* Ledeb.), meadowsweet (*Filipendula ulmaria* (L.) Maxim.), great willowherb (*Epilobium hirsutum* L.) lavishly grow. Almost always one can find melancholy thistle (*Cirsium heterophyllum* (L.) Hill.), nettle (*Urtica dioica* L.), white nettle (*Lamium album* L.).

In the headwaters of Bordyanka the mire surface is open to boast of sporadic willows (*Salix cinerea* L.), birches (*Betula pubescens* Ehrh.) and pines (*Pinus sylvestris* L.). In the lower course of the river along its channel basket willow (*Salix viminalis* L.) appears and the entire surface of the mire is covered with a dense stand of birch trees.

On the slopes of the valley Bordyanka, which descends to the swamp, one can be find some plant communities typical for different landscape-geographical provinces of Western Siberia. The southern slopes are characterized by forest-steppe kind of vegetation. The slopes of the northern exposition may be associated with the taiga communities existing at different stages of the natural and anthropogenic successions.

**Table 1.** Rare and endangered species of plants in the study area.

| No. | Species name | No. | Species name |
|---|---|---|---|
| 1 | *Erythronium sibiricum* (Fisch.et May.) Kryl. [a, b] | 13 | *Anemone caerulea* DC. [c] |
| 2 | *Hemerocallis minor* L. [c] | 14 | *Pulsatilla patens* (L.) Mill. [c] |
| 3 | *Lilium pilosiusculum (Freyn) Miscz.* [c] | 15 | *Actaea erythrocarpa* Fisch. [c] |
| 4 | *Cypripedium guttatum Sw.* [c] | 16 | *Corydalis bracteata* (Steph,) Pers. |
| 5 | *Cypripedium guttatum Sw.* [c] | 17 | *Viola mirabilis* L. [c] |
| 6 | *Dactylorhiza fuchsii (Druce) Soó* [c] | 18 | *Daphne mezereum* L. [c] |
| 7 | *Lychnis chalcedonica L.* [c] | 19 | *Circea alpina* L. [c] |
| 8 | *Dianthus versicolor Fischer ex Link.* [c] | 20 | *Primula macrocalyx* Bunge |
| 9 | *Paeonia anomala L.* [c] | 21 | *Polemonium caeruleum* L. [c] |
| 10 | *Aconitum anthoroideum DC.* [b] | 22 | *Brunnera sibirica* Stev. [b] |
| 11 | *Adonis sibirica Patrin ex Ledeb.* [c] | 23 | *Myosotis krylovii* Serg. [c] |
| 12 | *Anemone altaica Fisch.* [c] | | |

[a] Species in the Red Book of the Russian Federation [8].
[b] Species in the Red Book of the Tomsk oblast [9].
[c] Species requiring local protection [10].

For instance, a southward right-bank slope of the river Bordyanka valley is mostly open, covered with meadow vegetation and some scarce caragana shrubs (*Caragana frutex* (L.) C. Koch). A dense cover of wild grass on the upper slope features such meadow-steppe species as leopard plant (*Ligularia glauca* (L.) Hoffm.), daylily (*Hemerocalis minor* L.), pasqueflower (*Pulsatilla patens* L.), dropwort (*Filipendula vulgaris* Moench.), salsify (*Tragopogon orientalis* L.). Around the slope there is a large abundance of martagon lily (*Lilium pilosiusculum* (Freyn) Miscz.). The main aspect in the flowering period is produced by Jacob's ladder (*Polemonium caeruleum* L.), northern wolfsbane (*Aconitum septentrionale* Koelle), various species of peavines. The species composition of herbaceous plants includes more than 60 species.

A thin birch forest (*Betula pendula* Roth) with a small trace of pine grows in the gutter strip and small ravines that dissect the slope. Occasionally, one can see Siberian peashrub (*Caragana arborescens* Lam.). On the aligned areas along the slope the grass stand is comprised of sedge (*Carex macroura* Meinsh.), bushgrass (*Calamagrostis epigeios* (L.) Roth) and pine purple grass (*C. Langsdorfii* (Link) Trin).

The left-bank slope of the valley is occupied by a mixed forest. The dominant species is birch there. There are also Siberian pine (*Pinus sibirica* Tourn.), spruce (*Picea obovata* Ledeb.) and fir (Abies sibirica Ledeb), which are typical representatives of the communities of the dark coniferous taiga, once widespread in the suburbs of Tomsk. Under the cover of dark coniferous species one can see a bed of short grass: wood sorrel (*Oxalis acetosella* L.), muskroot (*Adoxa moschatellina* L.), Circea alpina L., and others. There are many fallen trees and stumps covered with green mosses and ferns.

A quite rare, owing to limited suitable habitats in the suburbs of Tomsk, common spotted orchid (*Dactylorhiza fuchsii (Druce) Soó*) is associated with those communities. There are as well Siberian trout lily (*Erythronium sibiricum* (Kryl.) Fisch.), which is still abundant on the forested slopes of that area, and *Brunnera sibirica* Stev. – a tertiary relic on the Red Books of the Tomsk oblast [9]. However, the most replete with rare and decorative species are the communities found on the open southern slopes of the valley in contact with the open reed fen.

### *4.2. Particular characteristics of mineral and peat deposits*

In the valley of the river Bordyanka there formed a unique carbonate-rich deposits consisting of peat, travertine and peat-travertine mixture. In the headwaters of the valley, under the cover of abundant reed thickets and reed waste lies the wet black peat of a high degree of decomposition. When exposed to hydrochloric acid, the peat effervesces from the surface. At the depth of 30-40 cm there are spots inside the peat deposits that were cemented with new formations of travertine. The level of peat waters varies from 15-20 cm at the foot of the valley slopes to 0 cm in water holes and closer to the river bed.

The carbonate deposits of the river valley were geologically studied as a deposit of travertines and peat-travertine mixtures (figure 2). It was established that the average thickness of the deposits is about 3 m [11], while the peat beds and peat-travertine mixtures in the main deposit of carbonate rocks are of a subordinate significance, though their bedding conditions are quite diverse. Occasionally the peat forms the very surface of the mire, accumulating in cavities of the travertine foundation. Sometimes, the peat deposits are uncovered in the base of the carbonate formation lying on the surface of the underlying Paleogene clays of the mineral bed of the mire. Sometimes, peat beds lie between layers of travertine. In some spots of the valley the total thickness of peat and travertine deposits is quite massive. In particular, the thickness of the carbonate-rich peat that formed in a travertine depression within the reed fen in the headwaters of Bordyanka is over 6 m. The total thickness of the carbonate deposits in this area is up to 8 m.

The general waterlogging of the river Bordyanka valley and the accumulation of lime material brings to the leakage of the fissure-veins waters of the Paleozoic aquifer and the overlying aquifer of the Neogene age. The process of travertine formation is observed almost everywhere, encompassing the vast slopes of the river valley. Given the total area of the open reed fen, which occupies the extended bottom of the valley of only about 8 hectares, the traces of carbonate deposits can be seen across the area of over 30 hectares.

The mineralization of water in the studied samples varies from 370 mg/l to 580 mg/l, pH is 7,1-8,2. The content of calcium and magnesium in the river and mire waters is 10-20% lower than in the spring water (table 2). According to a number of determined indicators, the maximum concentrations were observed on the contrary in the river or peat waters. However, some additional studies are required in order to establish any regularities in the distribution of chemical elements in the local landscape-geochemical system.

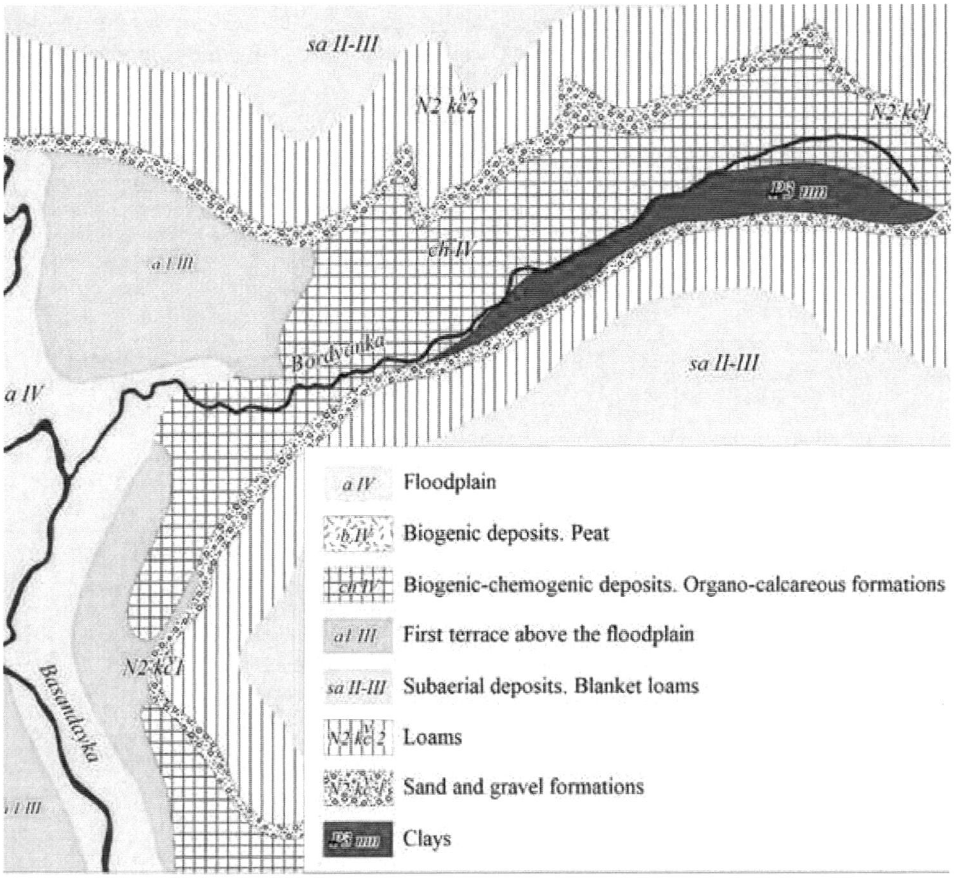


**Figure 2.** Features of the geological structure of the study area.

### *4.3. Prerequisites and forms of nature management*

The environmental conditions of the territory under study determine the presence of two types of natural resources, which have a traditional economic significance and are used in one way or another by the local population. Firstly, it is the underground waters that discharge to the surface on the slopes of the river Bordyanka valley by means of numerous springs. Secondly, it is the specific organo-mineral deposits lying at the floor of the valley, which belong to valuable agricultural raw materials that can be applied to increase fertility of acidic soils in the farmland zones of the Tomsk oblast.

The wedging out subsurface waters in the river Bordyanka valley have long been known as healing and widely used by the population. Not only the local people but also city residents come there from Tomsk to take the water from those springs for their own use. The water users are usually confident about the useful properties and quality of the spring water, though the sanitary conditions of the springs is not always satisfactory. Out of 33 of all known springs in the river valley only 2 are landscaped and arranged.

The balneological properties of the Paleozoic horizon water from the springs and wells of this area have been studied at the present time by the Tomsk Research Institute of Health resorts and Physiotherapy. According to the results, the water was qualified as natural mineral drinking water. It was found to contain sufficient quantities of silicic acid acting as a biologically active component of potable mineral waters. Its content varies from 23.4 to 36 mg/dm$^3$, while the balneological norm in the Russian Federation is 50,0 mg/dm$^3$. However, the water is suitable for treatment and prevention of gastrointestinal diseases. The active role of low concentrations of silicic acid for the biological effect of low-salt waters was experimentally substantiated and is taken into account when justifying the quality of health-related resources in a number of health resorts in the south of Western Siberia. The clinical studies of the local water have shown its unique properties: it reduces blood cholesterol and glucose and has a protective effect on gastric mucosa [12].

**Table 2.** Content of some chemical elements in water samples.

| Feature | Units | Wells at the headwaters in the river Bordyanka valley | Springs at the headwaters in the river Bordyanka valley | Open reed mire in the river Bordyalka valley | Bordyanka river downstream of the reed fen |
|---|---|---|---|---|---|
| K | mg/l | 0,60 | 0,64 | **0,94** | 0,53 |
| Na | | 5,62 | 5,82 | 5,64 | **6,73** |
| Mg | | 17,33 | **18,29** | 16,71 | 16,61 |
| Ca | | 111,04 | **122,79** | 109,34 | 102,42 |
| Fe | mkg/l | 5,6 | 14,2 | **181** | **141,7** |
| Mn | | 0,27 | 1,86 | 14,68 | **63,26** |
| Cu | | 0,21 | 0,18 | 0,39 | **1,47** |
| Zn | | 5,54 | 5,46 | **8,28** | **8,17** |
| Pb | | 0,074 | 0,067 | **0,171** | 0,066 |
| Ni | | 3,97 | 3,9 | **4,07** | 3,7 |
| Co | | 0,779 | 0,601 | **0,628** | 0,489 |
| V | | 0,862 | 0,618 | **1,17** | 0,322 |
| Mo | | 0,401 | 0,369 | **0,387** | 0,377 |
| Cd | | 0,002 | 0,002 | **0,003** | 0,001 |
| Cr | | 2,12 | **3,211** | 0,779 | 0,007 |
| Ag | | 0,024 | 0,058 | 0,046 | **0,067** |
| Be | | 0,010 | 0,017 | **0,035** | 0,022 |
| Sb | | 0,066 | 0,062 | 0,059 | 0,061 |

The useful properties of the groundwater from the springs and wells can be used to conduct health-related and recreational activities of the 'Klyuchi' Rehabilitation Centre located in the area, which will certainly increase its attractiveness as a health-improvement institution. The use of the spring waters with due account of interests of the local population is connected with solving the issues of providing public amenities and general landscape improvement in the places of their localization.

Whereas the use of groundwater from the springs can be made harmless provided the site is maintained and landscape is improved, the potential use of peat-travertine deposits in the river Bordyanka valley will result in a fundamental disturbance of the natural landscape. According to the data from the geological surveys, the deposits of agricultural raw materials in the river valley is 1 000 000 tons. It is the largest known deposit of this type in the Tomsk oblast. Given the $CaCO^3$ content of 75%, the reserves of calcareous material are 750 000 tons [11].

The high useful yield of carbonate material and its quality have long determined the interest in the development of this deposit. The sphere of application of carbonate raw material was considered in quite a broad manner that is from its use in agriculture to application in construction works. In particular, as early as in the 1940s of the last century it was believed that the deposits of calcareous tuff in the river Bordyanka valley is a material that allows, if properly burned, to obtain a high-quality lime. The conditions of occurrence and a small amount of surface stripping required allow the open-pit development of the deposit. Drainage appears to be easily achievable by excavation of the downstream carbonate sections [11].

However, it should be taken into account that the development of peat deposits of different quality and condition is generally connected with major disruptions of the environment. The development of peatlands and other types of wetlands in recent years has been conducted amid a backdrop of an increasing interest in their preservation in a natural state and a variety of conflict situations between the local economic needs and environmental initiatives [13]. The issue of rehabilitation of disturbed peatlands in spite of at least one hundred years of experience remains a topical environmental issue worldwide [14], especially when it comes to restoring the specific flora of peatlands and preservation of their bio-diversity [15]. In this regard, the lack of any positive experience in the peatlands reclamation in the Tomsk oblast raises concerns that the potential consequences of developing peat-travertine deposits in the river Bordyanka valley may become irreparable. Particular concerns are raised about undesirable consequences of a radical landscape transformation, in case of development of the deposit, and its close proximity to a large health improvement institution of Tomsk as well as the traditional use of the surrounding territories for recreation activities and construction of summer houses and suburban villages.

So the above may be used to support the effort for preservation of the natural landscape and traditional forms of the environmental management in the vicinity of the village Klyuchi in the southern suburbs of Tomsk, the restrictions on construction and further transformation of the territory, the implementation of its scientific and informational and educational potential, and support and development of its recreational functions through special improvement and organizational measures. These tasks may be solved through planning and organization of a specially protected multi-use area such as a landscape park.

## 5. Conclusions

The reed mire studied under this article is a part of a unique natural landscape of a small river valley in the southern suburbs of Tomsk. The mire stands in the way of housing and nature development thus aiding to preserve the biological and landscape diversity concentrated in this area.

The mire is a singular example of an intensive accumulation of carbonate-rich peat and travertine in the Tomsk oblast. The process of formation of peat-calcareous deposits in this area is of a high scientific and educational interest. Their economic development will result in some irreparable consequences.

The objectives of protection of the unique reed mire are consistent with the interests of preserving the traditional social and economic functions of the surrounding landscape and wise use of its recreation resources. The area of interest has long been used for recreation purposes. The exploitation of organic and mineral deposits of the mire as well as housing development in the adjacent areas will result in a decline or a total loss of aesthetic and recreational values of the area.

For the purpose of preservation of the reed mire and other natural and cultural monuments of the said area in the suburbs of Tomsk it is essential to establish a specially protected area under the patronage of regional authorities

## Acknowledgments

We acknowledge support from Tomsk State University (Competitiveness improvement Programme).

## References

[1] Liss O L, Abramova L I, Avetov I A, Berezina N A, Inisheva L I, Kurnishkova T V, Sluka Z A, Tolpysheva T Yu and Shvedchikova N K 2001 *Peatlands systems of Western Siberia and their environmental significance* ed V B Kuvaev (Tula: Grif and Co)

[2] *Peatlands of Western Siberia – their value in the biosphere* 1998 ed A A Zemtsov (Tomsk: Tomsk University Publishing)

[3] Semenova N M 2014 Western Siberia in the context of global nature conservation concerns *International Journal of Environmental Studies* **71** **(5)** pp 595–604

[4] Lapshina E D 2003 *The Flora of the Southeast of Western Siberia Peatlands* (Tomsk: Tomsk University Publishing)

[5] Semenova N M, Amelchenko V P and Volkova I I 2015 Botanical and geographical peculiarities and preservation prospects for a unique natural landscape located in the southern suburb of the city of Tomsk *Regional Environmental Issues* **6** 49–54

[6] Karandashev V K, Orlova T A, Lezhnev A E 2006 *Determination of the elemental composition of natural and drinking water by ICP-MS. Methods of quantitative chemical analysis*

IOP Conf. Series: Earth and Environmental Science **138** (2018) 012015 doi:10.1088/1755-1315/138/1/012015

(Moscow: Publishing house VIMS )

[7] Semenova N M 2001 The state, use and protection of peatlands resources in the Tomsk region *Nature protection: collection of articles* **2** ed Berezin A E (Tomsk: NTL Publishing) pp 69–86

[8] *The Red Book of the Russian Federation: Plants and Mushrooms* 2008 ed L V Bardunov and V S Novikov (Moskow: KMK Scientific Press)

[9] *The Red Book of Tomsk Region* 2013 ed A M Adam (Tomsk: Print Manufactory)

[10] *Rare and Endangered Species of Animals and Plants of the Tomsk Region* ed I P Laptev (Tomsk: Tomsk University Publishing)

[11] Makarov G Ya, Kovaleva L A 1980 Estimation of prospects and predicted reserves of freshwater calcareous raw materials in the zone of intensive agricultural development of the Tomsk region (Tomsk, Tomskaya geologorazvedochnaya ekspeditsiya Publ)

[12] Semenova N M, Nazarov A D, Loyko S V, Sidorina N G, Tishin P A 2017 Status and prospects of using he health improvement area «Klyuchi» (Tomsk region) *Bulletin of the Tomsk Polytechnic University. Geo Assets Engineering* **328 (9)** 127–139

[13] Cruickshank M, Tomlinson R, Bond D, Devine P and Edwards C 1995 Peat extraction, conservation and the rural economy in Northern Ireland *Applied Geography* **15 (4)** 365–383

[14] Parry L E, Holden J, Chapman P J 2014 Restoration of blanket peatlands *Journal of Environmental Management* **133** 193–205

[15] Woziwoda B, Kopeć D 2014 Afforestation or natural succession? Looking for the best way to manage abandoned cut-over peatlands for biodiversity conservation *Ecological Engineering* **63** 143–152

IOP Conf. Series: Earth and Environmental Science **138** (2018) 012016 doi:10.1088/1755-1315/138/1/012016

# Influence of north climatic conditions on the peat lipids composition

**O V Serebrennikova [1, 2], E B Strelnikova[1], M A Duchko*[2], Yu I Preis[3]**

[1] Institute of Petroleum Chemistry SB RAS, 4 Akademichesky st., Tomsk, 634021, Russia
[2] National Research Tomsk Polytechnic University, 30 Lenina st., Tomsk, 634050, Russia
[3] Institute of Monitoring of Climatic and Ecological System SB RAS, 10/3 Akademichesky st., Tomsk, 634021, Russia

*Email: maria.duchko@gmail.com

**Abstract**. The paper studies the composition of lipid organic compounds of peat from the northern regions of the Russian Federation. Peat was sampled in the northern taiga, forest-tundra and tundra zones, characterized by various hydrothermal conditions and vegetation cover. n-Alkanes, fatty acids and their ethers, aldehydes, ketones, alcohols, tocopherols, squalene, bi-, tri- and pentacyclic terpenoids, as well as steroids were identified in peat lipids by gas chromatography-mass spectrometry. The dependences of the total content of lipids and the majority of the investigated compounds classes on the ambient temperature and vegetation, as well as the correlation between the composition of n-alkanes and humidity were revealed.

## 1. Introduction
Hydrothermal conditions are very important in the process of peat formation [1]. The accumulation of green phytomass depends directly on the amount of heat and the duration of the vegetation period [2]. Therefore, an increase in the temperature of the environment will facilitate the deposition and transformation of a larger mass of peat-forming plants into the peat. Sharp cooling leads to a change in the vegetation cover and, as a consequence, to a change in the species composition of plant communities and the products of plant transformation in the process of peat formation. As a result, the composition of peat lipids should respond to changes in temperature conditions. Drying of mire in the result of decreasing atmospheric precipitation causes an increase in the growth and activity of aerobic bacteria, leading to lowering in the content of lipid components in the peat [3].

Understanding the mechanisms that unite vegetation and climate is one of the most important fundamental problems in paleoecological and paleoclimatic studies. It is very important for predicting the future state of the biosphere and assessing the impact of anthropogenic factors on it that may lead to climate change [4]. The peat from the northern taiga, forest-tundra and tundra zones, characterized by different hydrothermal conditions and vegetation cover, was studied to determine the influence of the climatic factor on the composition of peat lipids in the northern regions of the Russian Federation.

## 2. Samples and analysis
The figure 1 shows the location of the sampling areas.

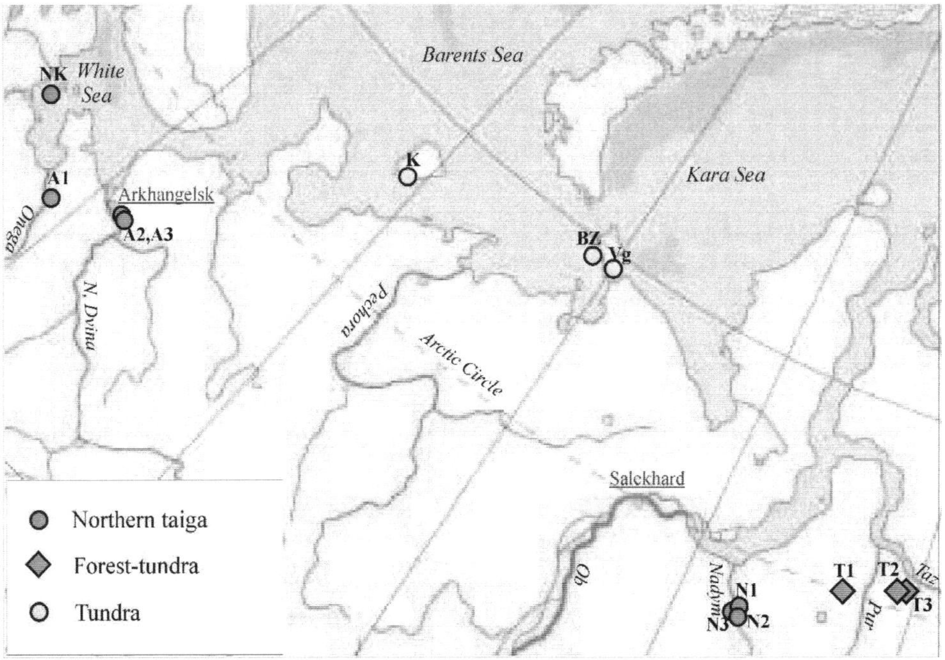


**Figure 1.** Map of Russian Arctic and location of study sites (see table 1 for identification of site index).

**Table 1.** Identification of the site index*** (* Fuscum peat, ** Sphagnum peat of bog hollows, *** All peat samples were taken from the depth of 0-0.2 m).

| Index | Region / District | T, °C | Average precipitation, mm |
|---|---|---|---|
| A1 | Arkhangelsk | 1.5 | 575 |
| A2* | Arkhangelsk | 0.8 | 558 |
| A3 | Arkhangelsk | 0.8 | 558 |
| N1** | Nadymskiy | -6.6 | 449 |
| N2 | Nadymskiy | -6.6 | 449 |
| N3 | Nadymskiy | -6.6 | 449 |
| T1 | Tasovskiy | -7.0 | 482 |
| T2* | Tasovskiy | -8.2 | 458 |
| T3** | Tasovskiy | -8.2 | 458 |
| | | Islands: | |
| NK | Nemecky Kuzov | 0.8 | 445 |
| K | Kolguev | -2.7 | 344 |
| BZ | Bolshoy Zincovy | -6.3 | 359 |
| Vg | Vaygach | -6.3 | 359 |

The studied highbog peat samples from the northern taiga were selected within the Arkhangelsk region and on one of the islands of the White Sea, where the current average annual temperatures range from 1.5°C to 0.8°C (table 1). The average annual precipitation is 575-558 mm in the continental part and 445 mm on the island Nemecky Kuzov. In the investigated section of the northern taiga subzone of Western Siberia the temperature is much lower, the amount of precipitation is close to one on the island Nemecky Kuzov. Among these samples there is one of bog hollows peat and the peat from arid areas. The investigated area of the forest-tundra zone (Western Siberia) is characterized by the lowest annual temperatures (among the studied areas) (table 2), the amount of precipitation is 458-482 mm. The samples also include peat located within swamps, as well as on elevated areas. They

IOP Conf. Series: Earth and Environmental Science **138** (2018) 012016 doi:10.1088/1755-1315/138/1/012016

are characterized by different amounts of precipitation and by different air humidity. The peat from tundra subzone was selected on three islands of the Barents Sea. The average long-term temperature at the Island Kolguev (northern tundra) is -2.7° C, the amount of precipitation is low. The amount of precipitation is higher in the arctic tundra on island Bolshoj Cinkovy and in the south of island Vaigach, where the average long-term temperature is about -6.3°C [2]. A detailed description of selection and preparation conditions and peat analysis is given in [5]. The data on the composition of n-alkanes were used to calculate the moisture index: Paq = $(C_{23}+C_{25})/(C_{23}+C_{25}+C_{29}+C_{31})$ [6]. It provides an approximate measure of the sedimentary input from submerged/floating aquatic macrophytes relative to that from emergent and terrestrial species.

## 3. Results and discussion

Acyclic alkanes, fatty acids and their ethers, ketones, alcohols, aldehydes and squalene, as well as bicyclic tocopherols and sesquiterpenes (BCT), tricyclic diterpenes (TDT), tetracyclic steroids and pentacyclic triterpenoids (PCT) were identified in the peat lipids (figure 2).

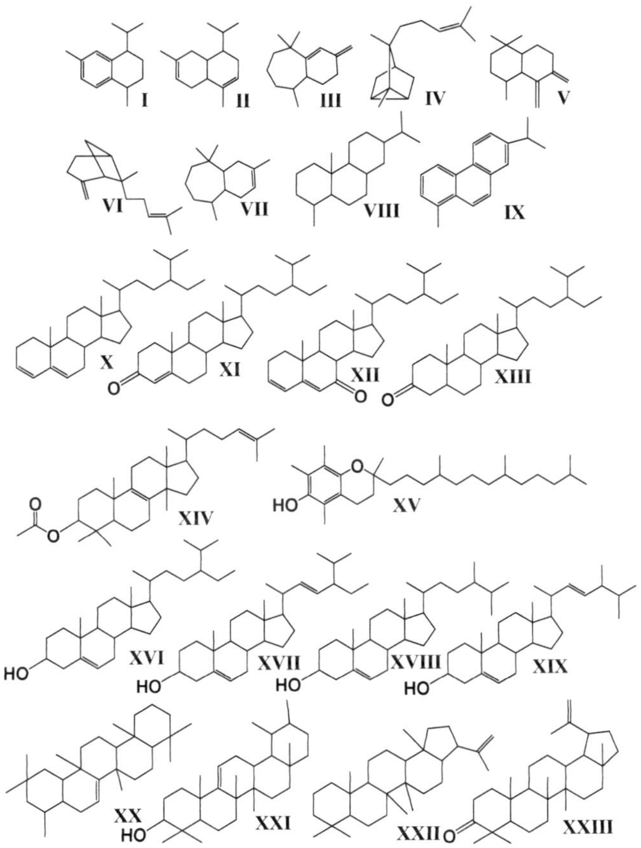


**Figure 2.** Major structural varieties of cyclic compounds in the peat lipids composition: I-calacorene, II-δ-cadinene, III-himachala-3(12),4-dien, IV-santalene, V-1,1,4α-trimethyl-5,6-dimethylene-decahydronaphtalene, VI-bergamotene, VII-himachalen, VIII-18-norabietane, IX-retene, X-stigmast-3,5-dien, XI- stigmast-4-en-3-one, XII-stigmast-3,5-dien-7-one, XIII-stigmastane-3-one, XIV-lanosterol acetate, XV-α-tocopherole, XVI- sitosterol, XVII-stigmasterol, XVIII-campesterol, XIX-crinosterol, XX-D-friedoolean-14-ene, XXI-ursen-3-ol, XXII-diploptene, XXIII-lup-20(29)-en-3-one.

**Table 2.** Group composition of the peat lipids from north taiga subzone.

| Pattern index | A1 | A2 | AP | N2 | N1* | N2 | N3 |
|---|---|---|---|---|---|---|---|
| North latitude | 63°82' | 64°20' | 64°19' | 64°57' | | 65°18' | |
| Eastern longitude | 38°49' | 40°37' | 40°41' | 35°10' | | 72°53' | |
| Lipid groups | | | | Content, mcg/g of dry peat | | | |
| n-Alkanes | 32.6 | 17.4 | 62.5 | 90.0 | 47.1 | 130.6 | 45.1 |
| Fatty acids | 15.6 | 4.9 | 5.4 | 8.8 | 9.3 | 10.3 | 9.5 |
| Fatty acids esters | 2.4 | 0.4 | 0.6 | 1.4 | 1.3 | 2.0 | 1.5 |
| Ketones | 16.7 | 2.8 | 5.1 | 10.8 | 1.6 | 18.7 | 11.4 |
| i-Alcohols | 1.8 | 1.1 | 1.1 | 0.7 | 7.7 | 1.5 | 1.0 |
| n-Aldehydes | 12.3 | 2.7 | 6.1 | 47.1 | 2.1 | 20.5 | 2.7 |
| Squalene | 4.6 | 1.5 | 0.9 | 4.0 | 1.3 | 0.7 | 1.4 |
| Tocopherols | 0.5 | 3.7 | 4.0 | 1.1 | 0.7 | 2.6 | 10.2 |
| Bicyclic terpenoids | 6.2 | 0.3 | 0.6 | 1.9 | 0.05 | 0.7 | 0.1 |
| Tricyclic terpenoids | 1.9 | 0.1 | 2.2 | 1.3 | 0.02 | 1.4 | 0.4 |
| Pentacyclic terpenoids | 12.8 | 76.3 | 50.4 | 21.6 | 0.9 | 18.8 | 17.5 |
| Steroids | 4.6 | 6.5 | 7.1 | 5.9 | 8.0 | 16.1 | 15.0 |

There is no dependence between the total lipid content in the peat, the temperature and the humidity of the environment. At the same time the average content of the peat lipids clearly reduces from the northern taiga to the tundra natural climatic zone (figure 3). It indicates a predominant influence of characteristic plant communities of natural zones on the content of peat lipids.

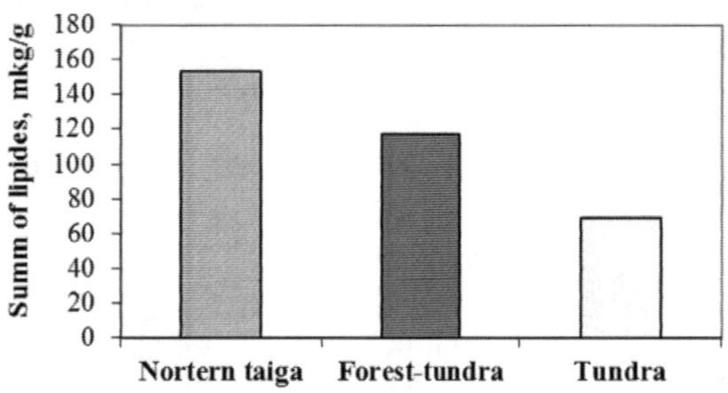


**Figure 3.** Average lipid content in the peat from various zones.

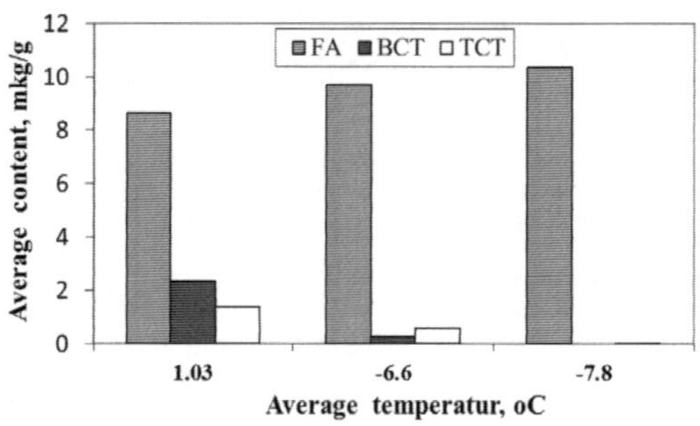


**Figure 4.** Average content of fatty acids (FA), BCT and TCT versus temperature of site.

The average content of fatty acids in continental peat samples with a constant individual composition slightly increases with decreasing the ambient temperatures (figure 4). In the peat sampled on tundra islands, the content of these compounds is much lower than in the other samples. The maximum content of fatty acids, as well as squalene and ketones, is fixed in the peat from the northern taiga in the area of the maximal temperature and amount of precipitation among the studied samples. The increased content of alcohols is fixed in the bog hollows peat. The content of aldehydes sharply increases in the peat from islands Nemeckij Kuzov and especially Kolguev, where aldehydes dominate among other classes of lipids. Such a high content of aldehydes can be a consequence of the participation of excrements of numerous nesting migratory birds in the formation of peat lipids [7]. It is confirmed by the predominance of 5β-isomers among cholestanones in the peat from the island Kolguev.

The content of acyclic ketones in the peat decreases from the north taiga to the forest-tundra and is minimal in the bog hollow peat.

The BCTs content in the northern taiga peat sharply decreases from the zone of positive ambient temperatures to the zone of negative ones (figure 4). A similar change is observed for TCTs. The content of the other classes of lipid organic compounds in the peat does not depend on the ambient temperatures.

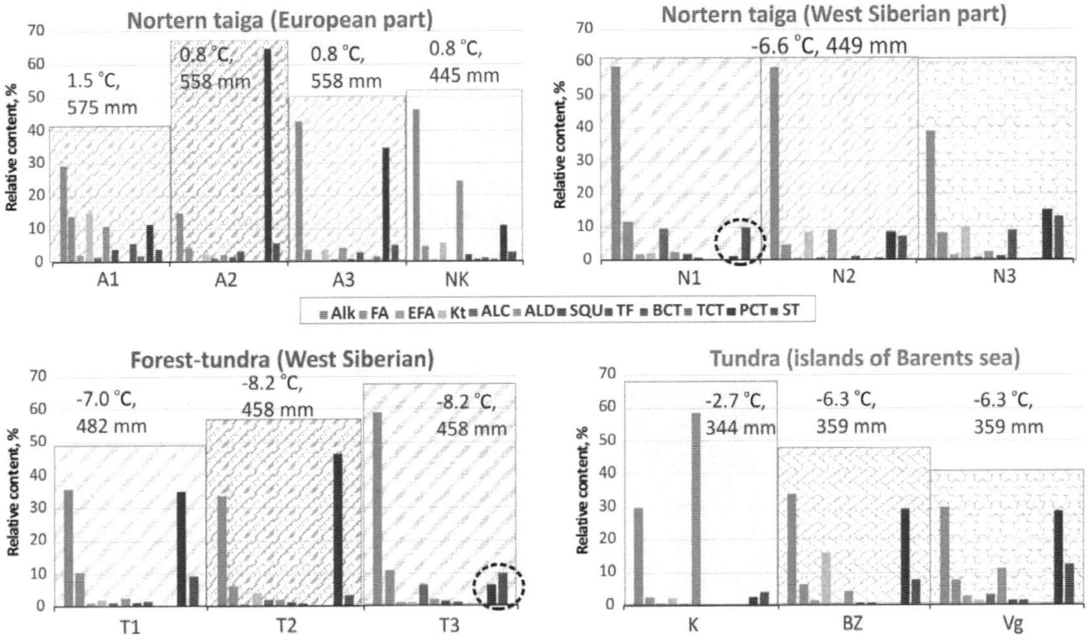


**Figure 5.** Relative content of separate groups of compounds in the peat lipids from the northern taiga, forest-tundra and tundra: Alk – n-alkanes, FA – fatty acids, EFA – esters of fatty acids, Kt – ketones, ALC – alcohols, ALD – aldehydes, SQU – squalene, TF – tocopherols, BCT – bicyclic sesquiterpenes, TCT – tricyclic diterpenes, PCT – pentacyclic triterpenoids, ST – steroids.

In the most part of the studied samples n-alkanes predominate in the peat lipids. PCTs predominate in the fuscum peat samples (A-2, T-2) regardless of the environment temperature and humidity. The bog hollows peat is characterized by the predominance of steroids over PCTs and the increased proportion of acyclic alcohols among the peat lipids, which present in trace amounts in other samples (figure 5).

**Table 3.** Group composition of the peat lipids from forest-tundra and tundra zones.

| Pattern index | T1 | T2 | T3* | K | BZ | Vg |
|---|---|---|---|---|---|---|
| North latitude | 66°19' | 67°20''' | | 69°01′ | 69°52' | 69°43' |
| Eastern longitude | 76°54' | 78°45'' | | 49°22′ | 59°27' | 60°03' |
| Lipid groups | | | Content, mcg/g of dry peat | | | |
| n-Alkanes | 30.1 | 46.2 | 74.2 | 36.4 | 17.6 | 9.6 |
| Fatty acids | 8.8 | 8.3 | 14.0 | 3.0 | 3.4 | 2.5 |
| Fatty acids esters | 1.0 | 1.2 | 1.7 | 0.9 | 0.7 | 0.9 |
| Ketones | 1.5 | 5.6 | 1.5 | 2.7 | 8.2 | 0.4 |
| i-Alcohols | 1.1 | 2.8 | 8.0 | 0.4 | 0.2 | 1.0 |
| n-Aldehydes | 2.2 | 2.6 | 2.6 | 72.2 | 2.3 | 3.7 |
| Squalene | 0.9 | 1.5 | 1.8 | 0.1 | 0.4 | 0.5 |
| Tocopherols | 1.2 | 1.0 | 1.3 | 0.1 | 0.3 | 0.5 |
| Bicyclic terpenoids | 0 | 0 | 0 | 0 | 0 | 0.03 |
| Tricyclic terpenoids | 0.005 | 0.01 | 0.1 | 0.03 | 0.01 | 0.1 |

IOP Conf. Series: Earth and Environmental Science **138** (2018) 012016 doi:10.1088/1755-1315/138/1/012016

| | | | | | | |
|---|---|---|---|---|---|---|
| Pentacyclic terpenoids | 29.5 | 63.3 | 8.3 | 3.1 | 15.2 | 9.3 |
| Steroids | 7.9 | 4.4 | 12.6 | 4.7 | 4.0 | 4.0 |

The high content of the $C_{23}$ homologue among n-alkanes was noted in the peat of the bog hollows and in the areas with the elevated amount of atmospheric precipitation (figure 6). This compound corresponds to moisture-loving plants [8]. In the other studied samples from the zones with a reduced amount of atmospheric precipitation, the content of $C_{23}$ n-alkane is not so high. Some odd n-alkanes ($C_{25}$-$C_{31}$) predominate in these samples. It corresponds, for example, to cotton grass predominated by $C_{31}$ n-alkane, to some species of sedges and horsetail, characterized by the dominance of $C_{29}$, to green moss and woody plants with a predominance of $C_{27}$ [8-11].

The value of Paq moisture index [6] decreases with a decrease in humidity from 0.39 to 0.15 for the peat of northern taiga lying at positive temperatures. In the mainland peat from the northern taiga and forest tundra, lying at a negative temperature, Paq decreases from 0.65-0.71 to 0.48-0.32; the maximum value (0.79) is recorded in the bog hollows peat. In the peat sampled on the island, the decrease in humidity leads to a decrease in the value of Paq from 0.50-0.61 to 0.13.

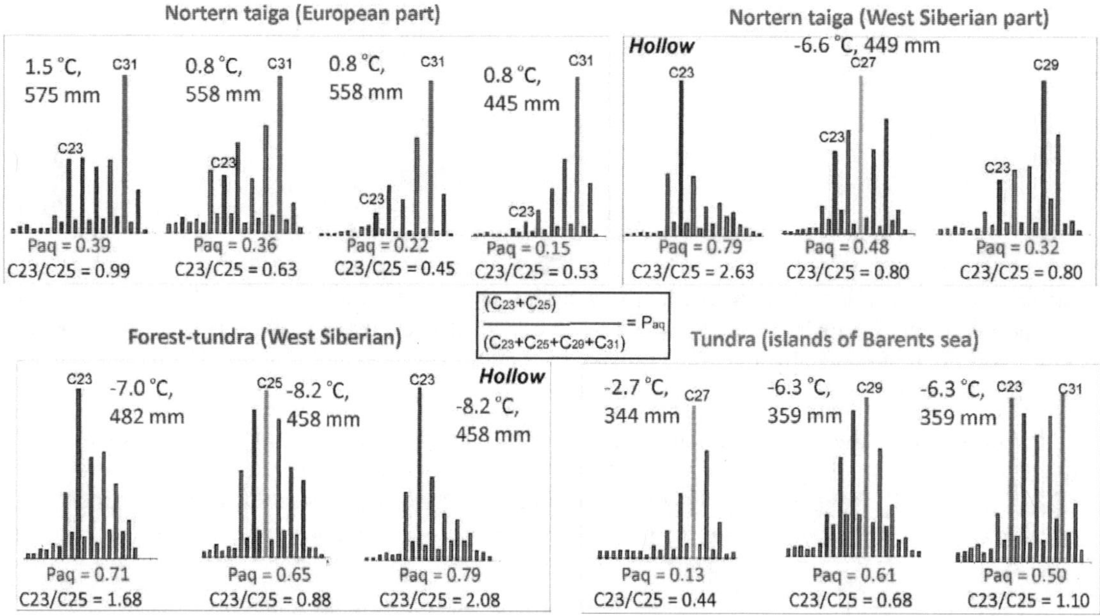


**Figure 6.** Molecular-mass distribution of n-alkanes in peat samples from the northern taiga, forest-tundra and tundra.

The ratio of moisture-loving sphagnum mosses ($C_{23}$ predominates) to sphagnum mosses of arid areas ($C_{25}$ predominates) [12] can be estimated by the ratio $C_{23}/C_{25}$ n-alkanes. The changes of the Paq values and the proportion of moisture-loving sphagnum mosses residues in the peat ($C_{23}/(C_{25}+C_{23})$) has a similar tendency with increasing amount of atmospheric precipitation for the most part of the peat samples (figures 7, 8).

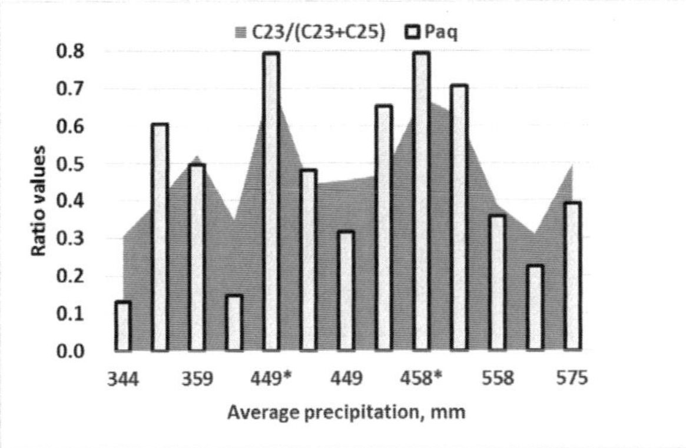


**Figure 7.** Influence of the amount of atmospheric precipitation on the value of the moisture index and the proportion of moisture-loving sphagnum mosses in the peat composition.

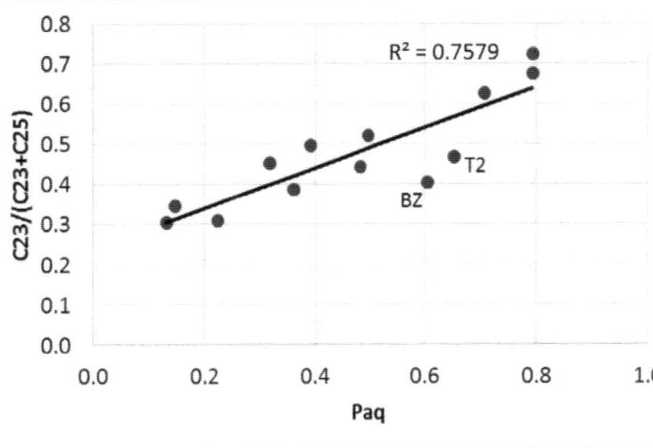


**Figure 8.** Correlation of moisture index and the proportion of moisture-loving sphagnum mosses in the peat composition.

The peat from the island Bolshoy Zinkovy is characterized by the increased Paq value. It may be explained by the limited distribution of vascular plants and the participation of their residues in the peat formation on this island. The discrepancy between Paq and the $C_{23}$ proportion in the fuscum peat T2 is caused by the absence of moisture-loving sphagnum mosses residues in the peat. It results in a low value of the ratio $C_{23}/(C_{23}+ C_{25})$. If these peat samples are excluded from the data array, the coefficient correlation between the Paq values and the n-alkane $C_{23}$ proportion increases to 0.93.

The figure 7 shows that the amount of atmospheric precipitation is not decisive in the formation of n-alkanes composition in the peat of the subarctic areas. The n-alkanes composition parameters proposed by (Ficken, 2000) for characterising the moisture reflect the intensity of atmospheric precipitation only within the limited areas. They directly depend on water saturation degree of the peat deposit surface and sharply increase in the bog hollows peat.

Calacorene and δ-cadinene predominate among BCTs in the peat at the temperature of +15°C, δ-cadinene and himachal-3(12), 4-diene – at the temperature of +0.8°C, δ-cadinene and santalene – at the temperature of -6.6 ° C. 1,1,4α-trimethyl-5,6-dimethylenedecahydronaphthalene predominates in the bog hollows peat. Further slight decrease in the BCTs concentrations is observed with a decrease of temperature in the forest-tundra. The main compounds among them are bergamotene and himachalene. In the bog hollows peat the main BCTs is 1,1,4α-trimethyl-5,6-dimethylene-decahydronaphthalene. In the tundra peat BCTs are presented only in the south of the island Vaigach. The predominant compounds among them are santalene and bergamotene.

In the peat from the positive temperatures zone TCTs are presented by decarboxylated derivatives of abietic and dehydroabietic acids with predominance of 18-norabietan in the most part of the samples. TCTs are absent in the peat from the island Bolshoy Zinkovy. In the peat from negative temperature zones they are represented by retene in low concentration.

The investigated peat steroids are presented by stigmast-3,5-dien and also by $C_{27}$-$C_{31}$ structures with a ketonic, alcohol or acetate substituent. The total concentration of steroids $C_{29}$ is maximum in all samples (figure 9). Among them prevails the group of ketones (dominated by stigmast-4-en-3-one, stigmasta- 5,7-dien-3-one or stigmastan-3-one). The concentration of 24-methylenecycloartanone ($C_{31}$) in the peat is not high. Lanosterol acetate predominates among the steroids in the shrub-sphagnum peat (N3) from north taiga, which differs from the others by the increased content of tocopherols (figure 5). Sitosterol and stigmasterol predominate among the steroids in the forest-tundra zone in the area with high humidity (T1) and in the bog hollows peat (T3) respectively. Saturated stigmastanones prevail in the peat from the island Kolguev with minimal amount of precipitation.

IOP Conf. Series: Earth and Environmental Science **138** (2018) 012016 doi:10.1088/1755-1315/138/1/012016

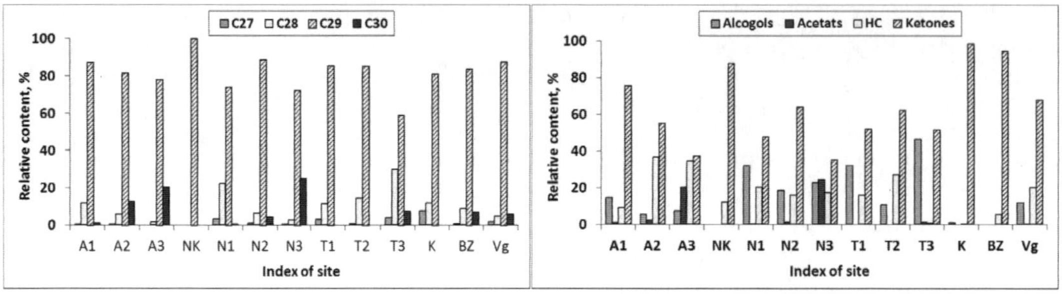


**Figure 9.** Relative amount of steroid groups in the peat lipids.

The bog hollows peat samples differ from the others in the increased relative content of $C_{28}$ steroids (figure 9), with the predominance of campesterol and crinosterol and steroid alcohols in general. The peat samples from the areas with a low amount of precipitation are characterized by the minimum proportion in the composition of steroids in the compounds with an alcohol group or their absence. A significant amount of lanosterol derivatives was detected only in the peat from the northern taiga. The amount of these compounds does not depend on the temperature and humidity of the environment and is determined by the plant communities of this natural zone. In particular, in the investigated dwarf birch *Betula nana* lanosterol occupies an appreciable position in the composition of steroids.

PCTs are presented by perhydropicene structures that predominate in most part of the studied peat samples and include derivatives of oleanene, ursene, hopanoids and lupene.

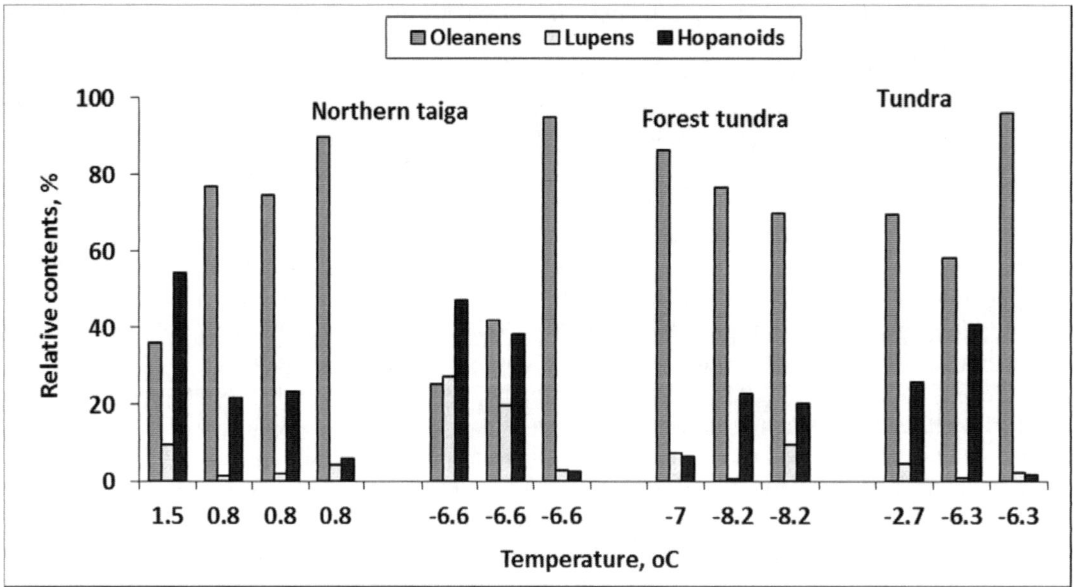


**Figure 10.** Relative amount of PCTs groups in the peat lipids.

Lupene derivatives are presented in a subordinate amount. Fuscum peat samples are characterized by the minimum content of these compounds regardless of the temperature and humidity of the environment. There is no dependence of the PCT composition on the climatic conditions and the water regime in the deposits. A high content of hopanoids is noticeable. It exceeds the concentration of perhydropicenes in the peat A1 sampled in the region with the maximum temperature and the amount of atmospheric precipitation.

IOP Conf. Series: Earth and Environmental Science **138** (2018) 012016 doi:10.1088/1755-1315/138/1/012016

## 4. Conclusion

Regardless of the ambient temperatures the bog hollows peat differs from the one sampled from the dry areas by the dominance of $C_{23}$ among n-alkanes, 1,1,4α-trimethyl-5,6-dimethylene-decahydronaphthalene among sesquiterpenoids, the increased proportion of campesterol derivatives and compounds with alcohol substituents in the composition of steroids. They are also characterized by the predominance of steroids over pentacyclic triterpenoids, the increased proportion of acyclic alcohols, and by low content of acyclic ketones. Fuscum peat samples differ in the predominance of pentacyclic triterpenoids among the peat lipids.

The peat samples from the areas with reduced precipitation are characterized by the minimum proportion of steroid compounds with an alcohol group or their absence.

The moisture index and the ratio of $C_{23}$ and $C_{25}$ n-alkanes, previously proposed for estimating the moisture level in the process of peat formation, reflect the intensity of atmospheric precipitation only within restricted areas.

The average content of the peat lipids regardless of the environment temperature and humidity decreases from the northern taiga to the tundra natural and climatic zones, indicating a predominant effect of characteristic plant communities of individual zones.

The decrease of the ambient temperatures in the continental peat samples leads to the slightly increase of the content of fatty acids. The concentrations of bicyclic sesquiterpenoids and tricyclic diterpenoids sharply decrease from the zone of positive temperatures to the zone of negative ones. Among sesquiterpenoids the content of calacorene, δ- and γ-cadinene reduces, santalene and bergamotene appear. Diterpenoids in the peat from the zone of positive temperatures are represented by decarboxylated derivatives of abietic and dehydroabietic acids with the predominance of 18-norabietan in the most part of the samples, the samples from the negative temperature zone – only by retene.

## References

[1] Andersson R A and Meyers P 2012 *Organic Geochemistry* **53** p 63
[2] Lavrinenko I A and Lavrinenko O V 2013 Proceedings of the Karelian Research Center RAS **6** p 4
[3] Svarovskaya L I, Serebrennikova O V, Duchko M A, Strel'nikova E B and Russkikh I V 2017 *Solid Fuel Chemistry* **51** p 67
[4] Brocks J J and Summons R E 2003 *Treatise on Geochemistry* **8** p 63
[5] Serebrennikova O V, Strel'nikova E B, Russkikh I V and Duchko M A 2017 *Solid Fuel Chemistry* **51** p 195
[6] Ficken K J, Li B, Swain D L and Eglinton G 2000 *Organic Geochemistry* **31** p 745
[7] Cooperband M F, McElfresh J S, Millar J G and Cardé R T 2008 *Journal of Insect Physiology* **54** p 1184
[8] Andersson R A, Kuhry P, Meyers P, Zebühr Y, Crill P and Mörth M 2011 *Organic Geochemistry* **42** p 1065
[9] Bush R T and McInerney F A 2013 *Geochimica et Cosmochimica Acta* **117** p 161
[10] Serebrennikova O V, Gulaya E V, Strel'nikova E B, Kaduchagov P B, Preis Yu I and Duchko M A 2014 *Khimija Rastitel'nogo Syr'ja* **1** p 257
[11] Serebrennikova O V, Strel'nikova E B, Russkikh I V, Preis Yu I and Duchko M A 2016 *Khimija Rastitel'nogo Syr'ja* **1** p 63
[12] Pancost R D, Baas M, van Geel B and Damsté J S S 2002 *Organic Geochemistry* **33** p 675

IOP Conf. Series: Earth and Environmental Science **138** (2018) 012017 doi:10.1088/1755-1315/138/1/012017

# Taxonomic composition of phytoplankton in the Vakh River (Western Siberia)

**O N Skorobogatova**

Department of Ecology, Nizhnevartovsk State University, Nizhnevartovsk, 56 Lenina Street, 628605, Russia

eco@nvsu.ru

**Abstract**. This paper provides data on the hydrological and hydrochemical parameters of the Vakh River in the Middle Ob region. In 2005-2008 we have identified 404 taxonomic units represented by 463 species, types and forms of algae, belonging to 140 genera, 52 families, 13 classes and 7 divisions. 386 species were identified for the first time, 141 taxa were identified as rare and 22 taxa as new for Western Siberia. Leading divisions, *Bacillariophyta* and *Chlorophyta*, make up for 78.9% of total phytoplankton diversity. *Cyanobacteria, Chrysophyta* and *Euglenophyta* form a community of 88 taxonomic units having a rank lower than genus level, and represent 19.0% of the total number. The floral role of *Xanthophyta* and *Dinophyta* is insignificant (2.1%). The main structure-forming species are 14: vegetating throughout the year (*Aulacoseira italica, Asterionella formosa*), summer taxa (*Microcystis aeruginosa, Melosira varians, Aulacoseira granulata, Pandorina morum, Pediastrum boryanum, P. duplex, Lacunastrum gracillimum, Scenedesmus quadricauda*) and summer-autumn (*Microcystis pulverea, Tabellaria fenestrata, T. flocculosa, Mucidosphaerium pulchellum*). The Vakh River demonstrates specific characteristics of boreal flowing waterways. The northern composition is represented in a big rate of families (44.2%) and genera (109 77.9) with one to three species. The richest genera *Closterium* (31 taxa), *Eunotia* (27 taxa), *Pinnularia* (22 taxa), *Desmidium* (11 taxa) and the family *Desmidiaceae* (45 taxa) ensure diversity of phytoplankton. Ecological and geographical analysis demonstrates predominance of cosmopolitan algae (56.6%). Plankton represents 44.7% of all algae, oligogalobs - 78.8% and indifferent algae – 36.7%. Water meets the requirements for β-mezosaprobian pollution zone, class of satisfactory purity (III class).

## 1. Introduction

First findings about phytoplankton in the Vah River date back to 2001 (Naumenko, 2001). The findings rely on the materials collected in the estuary of the river (June 1981-1985 and July 1986) and present 105 species of algae belonging to 5 groups. Further researches conducted in the area of Middle Taiga Vakh River allowed us to identify more types of plankton, which we added to the list. Among newly discovered types is *Stelexomonas dichotomus* Lack (Safonova, Shaulo, 2006). The research for the phytoplankton in the Vakh River may be called fragmentary, however, it is still immensely valuable for one of the major oil producing regions. The main objective of the study is to assess floristic and saprobiological characteristics of phytoplankton in the Vakh River. Specific objectives are to establish diversity of plankton algae in the Vakh River and to give taxonomic, ecological, geographical and saprobiological characteristics to the phytoplankton in the area.

## 2. Materials and methods

### *2.1. Study area*
The Vakh River is a deep river in Western Siberia, a right-bank tributary of the Ob River, flowing in the forest area of Khanty-Mansiysk Autonomous Area located in latitudinal, East-West direction, roughly on the 61st parallel. The flatness of the area (along with height differences from 170 to 32 meters above the sea level) defines the River's tortuosity index of 3-4. The Vakh River is the mainstream river in the Nizhnevartovsk Region with asymmetric basin, mainly right-bank prevalence. The wetland areas in the river amount to 60-80%, the drainage area covers over 76 thousand square kilometers, the average annual runoff ranges from 411 to 632 cubic meters. The runoff mostly attributes to snow melting, which amounts to 65% of the annual runoff. Its 30% attributes to groundwater and 5% - to rain. The seasonal floods in the River basin last from 2 to 2.5 months; the water level during springtime rises up to 7.5 – 9.0 meters compared to wintertime and the peak is in the middle of June (Beirom, 1975). The climate in the watercourse basin is continental with brief interseason period; summer season lasts for 95 days (Beirom, 1975). The observations allow us to conclude the following: (1) the rivers remain completely frozen for 178-222 days; (2) the maximum ice thickness is 71 centimeters; (3) the snow height is 40 centimeters. The river shifts to the summer conditions in the first part of June. The water temperature reaches +21°C, occasionally it is +26°C by the end of July and it drops to +4°C in the last decade of October. The critical drop below +0.2°C is generally observed in the end of October – beginning of November.

The Vakh River is 964 kilometers (Beirom, 1975). The width of the river varies from just several meters near the river head and 600 meters around the estuary; its depth is around 19 meters, the speed is 0.3 – 1.1 m/s, the coastline height in some places is over 40 meters. According to its morphometric, hydrologic and hydrochemical characteristics, the river is divided into three areas – upstream, middlestream and downstream.

The chemical composition of the water is hydrocarbonated with mineralization reaching 13 - 88 mg/L. The water is soft, its hardness amounts to 1.5 mg/L (Alekin, 1953). The total amount of ions does not exceed 93.4 mg/L (Beirom, 1975). The Vakh River had demonstrated low water transparency during all the years of study (2005-2008). In the period between March and October it varies between 12 and 33 cm and pH value ranges from 5.3 to 8.1. The degree of peat formation determines high algal blooming and high iron amount. The Khanty-Mansiysk Regional Centre for Hydrometeorology and Environment Monitoring established that in the period of 2005-2007 there were high levels of algal blooming ranging 35 – 196 degrees, BOD was 0.1 – 1.8 mg/L and the amount of iron was 0.3 – 9 mg/L.

### *2.2. Collected material*
Seven hundred and twenty-three phytoplankton samples, collected along the Vakh River from 2005 to 2008 (during all seasons), were used for the study.

The phytoplankton was collected using routing method and stationary method. It was collected in one liter containers along the river (left bank, right bank and the middle of the river) at the depth of 20 cm. The process followed all the proper methods and requirements for sample material collection and processing (Fjodorov, 1979; Guidelines…1981; Algae, 1989). Other measurements needed for the study were taken: width, depth, speed of the water flow. A surface bucket thermometer was used for measuring the temperature on the surface. The transparency was estimated via Secchi disk; pH values were measured with a portable pH-tester "pHs can WP2".

The observations were conducted in the area of village Korliki and Laryak (upstream Vakh River), village Vakhovsk (middlestream Vakh River), and in the area of ferry crossing (downstream) during the year. The samples were taken three times per month in three places: right and left banks and centre of the streambed (Figure 1).

IOP Conf. Series: Earth and Environmental Science **138** (2018) 012017 doi:10.1088/1755-1315/138/1/012017

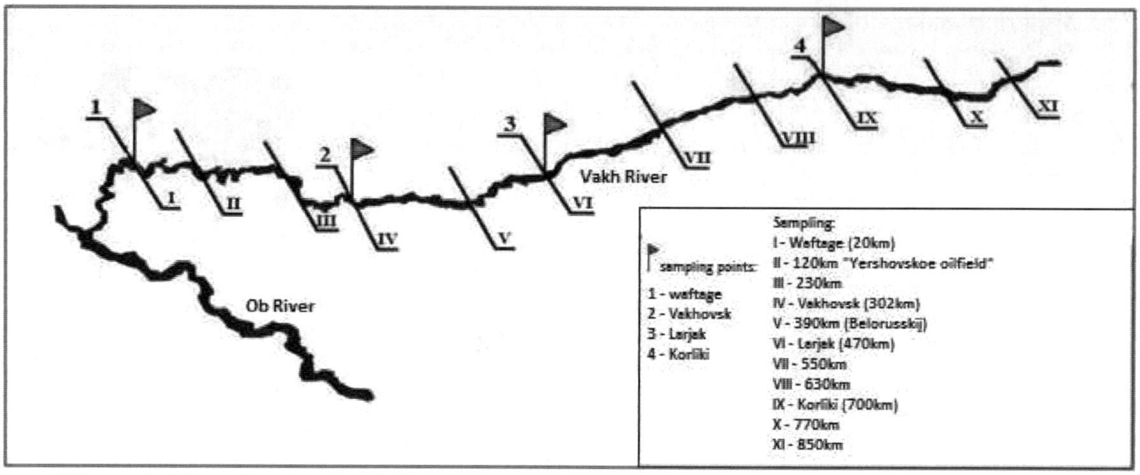


**Figure 1.** Map of stationary (1-4) and rout (I-XI) water sampling points on the Vakh River.

Sedimentation was used to determine the amount of algological material. The qualitative composition of the samples was determined using cone-shaped net and gauze No. 76.

All observations were performed on living or fixed in 4% solution of formalin material. All algae were studied under microscopes "Amplival" (Carl Zeiss, Jena) and «Mikmed-5» (Lomo, St. Petersburg) with manifold increase of 640 – 1600. We processed the samples to eliminate protoplast of cells (Zabelina, Kiselev et al., 1951; Algae, 1989) to establish what species the diatoms belong to. Cleaned diatom valves were mounted into Canada balsam with refractive index of 1.68 (Elyashev, 1957).

We referred to generally recognized national and foreign field guides and guide books on algology to identify the collected samples. In the systematic positioning of algae we accounted for changes and revisions for Cyanophyta (Kondratyeva, 1972; 1975; 1981; 1984), Chlorococcales (Tsarenko, 1990; 2005), *Pediastrum, Scenedesmus* (Hegemald, 2000) and Desmidiales (Palamar-Moedvintseva, 1982). The following field guides were used to identify diatoms (Gollerbakh…, 1951; Lange-Bertalot, 2001; Krammer.., 2000; 2002; 2003). Floristic composition analysis was conducted with the use of systems proposed by Russian diatomologists Glezer Z.I., Karaeva N.I., Makarova I.V. et al. (Diatoms…, 1988) with account for recent revisions of *Aulacoseira, Fragilaria, Diatoma, Achnanthes, Navicula, Pinnularia, Cymbella, Gomphonema, Eunotia* (Bukhtiarova, Round, 1996; Genka, Bondarenko, Schur…, 2011; Kharitonov, Genkal…, 2012, Dorofejuk, Kulikovsky…, 2012). "Biodiversity of algae indicators" field guide was used to conduct ecological analysis (Barinova et al., 2006). The data is presented in the tables (Table 1-4).

## 3. Results and discussions

We identified 404 taxonomic units represented by 463 species, varieties and forms of algae, belonging to 140 genera, 51 families, 13 classes and 7 divisions (Table 5). Chlorophyta has the biggest number of species - 40.1 %; it is followed by Bacillariophyta – 38.9%; Chrysophyta – 7.0%; Cyanophyta — 5.9%; Euglenophyta – 5.9%. Xanthophyta and Dinophyta comprise 2.2 % (Table 1).

In order to give a full and precise overview of the plankton algae in the Vakh River, the author assesses floristic characteristics at the level of infraspecific taxonomic units (further: IST) including type-species.

Chlorophyta and Bacillariophyta make up species composition of phytoplankton in the Vakh River and comprise 78.9% of total IST plankton. It is worth noting that it is not only Bacillariophyta that plays an important part in the composition of phytoplankton of the Vakh River, but also Chlorophyta. Their share is almost equal to 39.4% for Bacillariophyta and 39.5% for Chlorophyta. Bacillariophyta usually dominates over other algal divisions in Siberian rivers. This observation is confirmed by the composition of the Ob River (Naumenko, 1996) and its tributaries (Naumenko, 1985, 1991, 1998), Viluy (Remigaylo, 1995), Molodo (Gabyshev, 1999), Nizhnyaya Kolyma (Kopyrina, 2009) etc.

**Table 1.** Systematic composition of plankton algae in the Vakh River.

| Division | Number of species | Share, % | Number of specific and intraspecific taxonomic units | Share, % |
|---|---|---|---|---|
| Cyanophyta (Cyanobacteria) | 24 | 5.9 | 31 | 6.7 |
| Chrysophyta | 28 | 7.0 | 31 | 6.7 |
| Bacillariophyta | 157 | 38.9 | 182 | 39.4 |
| Euglenophyta | 24 | 5.9 | 26 | 5.6 |
| Xanthophyta | 7 | 1.7 | 7 | 1.5 |
| Dinophyta | 2 | 0.5 | 3 | 0.6 |
| Chlorophyta | 162 | 40.1 | 183 | 39.5 |
| Total: | 404 | 100 | 463 | 100 |

The first five classes in ranking were found to encompass 407 algae, which represent 87.9% of the total plankton found. The most numerous is Pennatophyceae with 171 IST. Its share in the total number of found algae is 36.9%. Pennatophyceae is followed by Chlorophyceae – 103 (22.2%), Conjugatopyceae – 76 (16.4%), Chrysophyceae – 31 (6.7%), Euglenophyceae – 26 (5.6 %). The amount of IST ranges from 1 to 17 in the remaining 8 classes.

The top ten families contain more than half of the identified algae taking the leading role in formation of algae in the River. 276 IST (57.3 %) were determined in Bacillariophyta and Chlorophyta leading families. It emphasizes their leading role in algae composition of the river. Euglenophyta and Chrysophyta are leading families and comprise 26 (8.2 %) and 14 (4.4 %) respectively, leading the research to the conclusion that these families are important for the phytoplankton of the river (Table 2).

**Table 2.** Leading families of phytoplankton of the Vakh River.

| Rank | Family | Number of specific and intraspecific taxonomic units | | Number of genera |
|---|---|---|---|---|
| | | absolute number | share, % | |
| 1 | Naviculaceae | 64 | 21.2 | 17 |
| 2 | Desmidiaceae | 45 | 14.9 | 13 |
| 3 | Scenedesmaceae | 43 | 14.2 | 8 |
| 4 | Closteriaceae | 31 | 10.3 | 1 |
| 5 | Eunotiaceae | 27 | 8.9 | 1 |
| 6 | Euglenaceae | 26 | 8.6 | 5 |
| 7 | Fragilariaceae | 21 | 7.0 | 6 |
| 8 | Selenastraceae | 16 | 5.3 | 7 |
| 9 | Surirellaceae | 15 | 5.0 | 3 |
| 10 | Dinobryaceae | 14 | 4.6 | 2 |
| | Total: | 302 | 65.5 | 63 |

Other families contain one third (166) of the identified algae which comprises 34.5%. They are presented in the descended order: Hydrodictyaceae (13 taxonomic units ranking lower than genus), Cymbellaceae (11), Chrysococcaceae (10), Microcystidaceae (8), Gomphonemataceae, Oocystaceae, Achnanthaceae, Nitzschiaceae (each 7), Anabaenaceae, Coelastraceae and Synuraceae (each 6), Stephanodiscaceae and Diatomaceae (each 5), Oscillatoriaceae, Aulacoseiraceae, Chlorellaceae, Tabellariaceae and Pleurochloridaceae (each 4).

The most represented are 12 genera that encompass 40.0% of the plankton algae (Table 3).

**Table 3.** Leading genera of phytoplankton of the Vakh River (based on the number of taxonomic units lower than genus).

| Rank | Genus | Taxonomic units lower than genus | % of the total number of algae |
|---|---|---|---|
| 1 | *Closterium* | 31 | 7.0 |
| 2 | *Eunotia* | 27 | 5.8 |
| 3 | *Pinnularia* | 22 | 4.8 |
| 4 | *Scenedesmus* | 17 | 3.6 |
| 5 | *Dinobryon* | 13 | 2.8 |
| 6 | *Surirella* | 12 | 2.5 |
| 7–9 | *Navicula* | 11 | 2.3 |
| 7–9 | *Desmodesmus* | 11 | 2.3 |
| 7–9 | *Staurastrum* | 11 | 2.3 |
| 10–12 | *Fragilaria* | 10 | 2.2 |
| 10–12 | *Trachelomonas* | 10 | 2.2 |
| 10–12 | *Cosmarium* | 10 | 2.2 |
| Total: | 12 | 185 | 40.0 |

Genera at the top of the list each contain from 31 to 10 taxonomic units ranking lower than genera. Genera are presented in the descending order: *Phacus*, *Kephyrion* – 9 IST each; *Pediastrum* – 8; *Microcystis* – 7; *Anabaena*, *Ulnaria*, *Gomphonema*, *Nitzschia* and *Staurodesmus* – 6 each; *Stauroneis* and *Tetrastrum* – 5 each; *Cyclotella*, *Aulacoseira*, *Planothidium*, *Cymbella*, *Cymbopleura*, *Monoraphidium*, *Coelastrum*, *Lagerheimia* – 4 each.

109 genera (77.9% of all genera) are species-poor. This entails that the largest part of the whole flora in the Vakh River is represented by small amounts of genera and families. It means that phytoplankton of the river is allochotonic and emphasizes the complexity of historical development of the species and plankton in the river.

In general, the distribution of families and genera in the phytoplankton of the Vakh River corresponds to the taxonomic spectrum of the northern flowing water bodies (Getsen, 1985; Vasilyeva, 1989; Vorobjeva, 1995; Naumenko, 1985, 1996).

There are twenty-five genera that are only present in one of the river areas and can be called specific to those areas: *Coelosphaerium*, *Nostoc*, *Rivularia*, *Plectonema*, *Chromulina*, *Mallomonas*, *Mayamaea*, *Achanthes*, *Rhoicosphenia*, *Pseudostaurastrum*, *Centritractus*, *Characiopsis*, *Katodinium*, *Ankyra*, *Hydrodictyon*, *Golenkiniopsis*, *Micractinium*, *Dictyosphaerium*, *Radiococcus*, *Franceia*, *Pseudodidimocystis*, *Pleurotaenium*, *Raphidiastrum*, *Pachyphorium* and *Bambusina*. Among the rare genera are ones that are known to be generally rare for Siberian water bodies: *Raphidiastrum*, *Golenkiniopsis*, *Micractinium*, *Franceia*, *Didymocystis* etc.

There are 91 genera (65% of all the genera present) develop in all parts of the river. It implies that the ecological conditions are similar along the whole river current in the latitudinal direction.

There are 190 taxonomic units (41.0 % of all) ranking lower genus similar for all the river parts, of them diatoms are 83, Chlorophyta – 77, Cyanobacteria – 13, Chrysophyta – 9, Euglenophyta – 6, Xanthophyceae and Dinophyta 1 each. The majority of these algae play an important role in structural composition of algae communities and are functional nucleus of the water stream. There are dominating species that develop all year around and maintain high population: *Aulacoseira italica*, *Asterionella formosa*, growing in the summer period *Microcystis aeruginosa*, *Melosira varians*, *Aulacoseira granulata*, *Pandorina morum*, *Pediastrum boryanum*, *P. biradiatum*, *Lacunastrum gracillimum*, *Scenedesmus quadricauda* and in summer – fall period: *Microcystis pulverea*, *Tabellaria fenestrata*, *T. flocculosa*, *Mucidosphaerium pulchellum* (Skorobogatova, 2012). *Aulacoseira italica* has the biggest coenotic value (Skorobogatova, 2010).

The abundance of species *Eunotia*, *Closterium*, *Pinnularia* and family Desmidiacea is a specific characteristic of the Vakh River and algal for the Taiga wetlands in Western Siberia. They develop in waters with low amount of salts and low acid reaction and form "boggy complex" which takes 27% of the whole composition. The valves of *Eunotia bilunaris*, *E. monodon*, *E. praerupta* and their variations, *E. undulata* and *E. neosibirica* are frequent in the samples. There are 9 representatives of these species that can be classified as rare (Naumenko, Skorobogatova, 2009). Representatives of *Closterium* species grow in the Vakh River from the first decade of June until middle of October with maximum species diversity appearing in the first decade of September. 5 taxonomic units of *Closterium* species that have not been determined earlier in the Middle Taiga were identified in the Vakh River phytoplankton (Skorobogatova, Naumenko, 2009). Desmidiales have a rich composition but are not evenly distributed along the river flow and do not have high populations.

The Vakh River flora proportions, ratio of number of genera and species to one family, is 1:2.7:9.1. The richness in genera along the stream is 3.3. There have been identified 141 specific taxonomic units; the coefficient of algae specificity of genera is relatively low – 17.9%, the specificity of species is considerably higher – 30.5%. The class of Conjugatophyceae demonstrates the highest specificity.

In the upstream of the Vakh River 255 species, types and forms of algae from 7 divisions, 41 families and 105 genera were found (Table 4).

**Table 4.** Systematic division of plankton algae in correlation with the River part.

| Division | Upstream | | Middlestream | | Downstream | |
|---|---|---|---|---|---|---|
| | Number of taxonomic units | % | Number of taxonomic units | % | Number of taxonomic units | % |
| Cyanophyta (Cyanobacteria) | 15 | 5.8 | 18 | 5.7 | 30 | 7.4 |
| Chrysophyta | 18 | 7.1 | 17 | 5.4 | 24 | 5.9 |
| Bacillariophyta | 106 | 41.6 | 129 | 41.0 | 165 | 40.4 |
| Euglenophyta | 7 | 2.7 | 11 | 3.5 | 25 | 6.1 |
| Xanthophyta | 4 | 1.6 | 3 | 0.9 | 4 | 1.0 |
| Dinophyta | 2 | 0.8 | 1 | 0.3 | 3 | 0.7 |
| Chlorophyta | 103 | 40.4 | 136 | 43.2 | 157 | 38.5 |
| Total: | 255 | 100 | 315 | 100 | 408 | 100 |

The leading families of the phytoplankton spectrum of the Vakh River upstream are Naviculaceae (34 specific and infraspecific taxonomic units), Scenedesmaceae (33), Eunotiaceae (21), Desmidiaceae (17), Closteriaceae (12), Dinobryaceae (11), Hydrodictyaceae (10). Five genera are distinguished based on the diversity of algae: *Eunotia* (21 species, types and forms), *Scenedesmus* (13), *Pinnularia* and *Closterium* (12 each), *Dinobryon* (10 IST). They encompass almost one third of the whole phytoplankton composition of the upstream Vakh River (26.7 %).

The coefficient of algae specificity of genera is 5.7%, the specificity of species is 7.8%. The flora proportions of the upstream area are 1:2.5:6.2; the richness of genera is 2.4.

The middle stream of the river hosts 315 species, types and forms made up of 110 genera, 50 families, 7 divisions. The top five families are Naviculaceae (48 taxonomic units ranking lower than genera), Scenedesmaceae and Desmidiaceae (33 each), Closteriaceae (25) and Eunotiaceae (23). In total, it makes up 51.4% of all identified. It is important to emphasize the leading position of the Chlorophyta (91 IST) and diatoms (71 IST). The leading genera of phytoplankton have 29.8% IST of all identified in the middle stream area: *Closterium* (24 IST), *Eunotia* (23), *Pinnularia* (16), *Scenedesmus* (13), *Dinobryon* and *Cosmarium* (9 each).

The coefficient of algae specificity of genera is 2.7%, the specificity of species is 8.9%. The phytoplankton proportions of the middlestream area is 1:2.2:6.3; the richness of genera is 2.8.

IOP Conf. Series: Earth and Environmental Science **138** (2018) 012017 doi:10.1088/1755-1315/138/1/012017

The estuarial part of the river houses 408 specific and infraspecific taxonomic units, 52 families and 124 genera, which are 56.7% of identified phytoplankton downstream Vakh River. The families are distributed as follows: Naviculaceae (59 IST), Scenedesmaceae (42), Desmidiaceae (34), Eunotiaceae and Euglenaceae (25 each), Closteriaceae (23), Fragilariaceae (21). The share of Diatoms in the leading families in the downstream Vakh River is 45.9%, Chlorophyta – 43.2%, Euglenophyta – 10.9%.

The most diverse in the genera composition in this area are *Eunotia* (25 IST), *Closterium* (23), *Pinnularia* (19), *Scenedesmus* (17), *Dinobryon* (12), *Navicula*, *Surirella*, *Desmodesmus* and *Trachelomonas* (10 each). The listed genera consist of 33.6% of all identified in the downstream Vakh River.

The coefficient of algae specificity of genera is 8.8%, the specificity of species is 23.3%. The flora proportions of the downstream are 1:2.4:7.9; the richness of genera is 3.3.

Therefore, the level of species diversity increases from the upstream areas to the estuary. The species, types and form diversity in the middle stream is of 60 representatives (19.0%) higher than in the upstream, the diversity in the downstream is 93 (22.8%) higher than in the middle stream and 153 (37.5%) higher than in the upstream.

Diatoms and Chlorophyta, which take top two positions, frequent lists of the leading algae for various water bodies and regions. Depending on water body type and its geographical location, they may vary in rankings and order.

Algae relations with main characteristics of the environment were accounted for in the study. Identified species are not homogenious in their ecological composition because the phytoplankton of the Vakh River is made up from benthos algae and fouling community, plankton algae of the catch basin and river plankton itself. Ecological analysis demonstrated that plankton algae have the highest diversity – 207 IST (44.7%). All identified plankton forms belong to seven divisions with leading representatives from Chlorophyta. Scenedesmaceae, Selenastraceae, Closteriaceae and others are noted for their diversity. *Microcystis aeruginosa*, *M. pulverea*, *Cyclotella meneghiniana*, *Aulacoseira distans*, *A. granulata*, *A. italica*, *Ulnaria acus*, *Asterionella formosa*, *Tabellaria fenestrata*, *Pandorina charkoviensis*, *Pediastrum boryanum*, *P. duplex*, *Actinastrum hantzschii*, *Acutodesmus acuminatus*, *S. serratus*, *S. quadricauda*, *Desmodesmus lefevrii* and others belong to the plankton that grows particularly intensively during one of the year periods. Appearance of *Merismopedia*, *Coelosphaerium*, *Rivularia*, *Oscillatoria*, *Chromulina*, *Mallomonas*, *Catenochrysis*, *Strombomonas*, *Lepocinclis*, *Cosmarium* is singular. Benthos algae is represented in 100IST (21.6%), of which the most common are *Navicula radiosa*, *N. viridula*, *Hippodonta capitata*, *Sellaphora rectangularis*, *Stauroneis anceps*, *Pinnularia gibba*, *Gyrosigma acuminatum* etc.

Periphyton includes 66 IST (14.3%) with more commonly appearing *Fragilaria construens*, *Ulnaria ulna*, *Eunotia bilunaris*, *E. monodon*, *E. undulata*, *E. praerupta*, *E. bidens*, *E. curtagrunowii*, *Cymbella naviculiformis*, *Eucyonema minutum*, *Gomphonema acuminatum*, *G. acuminatum* var. *coronatum*. Only one species of attache bioforms from Cyanobacteria was found. It is called *Lyngbya kuetzingii* and it was found only once downstream of the river.

The distribution of algae in downstream, middlestream and upstream area in correlation to the environmental conditions is similar to the distribution of algae in the river in general.

The algae of the River are divided into oligogalobs and mezogalobs. The oligogalobs include 365 IST (78.8%) with leading indifferent forms – 286 (61,7%), galofobs and galofils – 47 and 32 IST (10.2% and 6.9%) respectively. *Chrysococcus rufescens*, *Meridion circulare* var. *constrictum*, *Neidium iridis*, *Eunotia* representative are common for soft waters of the Vakh River. Only one species of mezogalobs, *Surirella ovalis*, was found. It was spotted in the downstream and upstream.

As for pH factor, neutral algae take the lead with 170 IST (36.7%), 50 algae (10.8%) are acidophile, which can be explained by fermenting from marshy flood bed. Aulacoseira distans, *Tabellaria fenestrata*, *T. flocculosa*, *Eunotia bilunaris*, *E. monodon*, *Ankistrodesmus falcatus* develop most actively.

The geographical distribution of algae populating a river is important in studies of phytoplankton. The majority of plankton is cosmopolitan, with taxonomic units ranking lower than genus (56.6%), and boreal – 71 (15.3%). Only 38 alpinoarctic taxonomic units ranking lower than genus (8.2%) have been identified. Stenothermal cold-loving diatoms *Cyclotella antiqua*, *Fragilariforma virescens*, *Meridion circulare*, *Eunotia praerupta*, rare alpinoarctic species *Tetracyclus lacustris*, *T. lacustris* var. *capitata*, *Stauroneis parvula*, *Frustulia rhomboides*, *Eunotia parallela* signify harsh weather conditions.

IST are markers of water saprobity. 126 of them belong to β-mesosaprobes, 62 to transitioning zone. Such tendency has been noted in all river parts. The richness and abundance of β-mesosaprobes and species of transitioning zone allows the Vakh River water to have III class of purity (satisfactory class).

**Table 5.** Systematic composition of phytoplankton in the Vakh River (U.S. – upstream river ; M.S. – middlestream river; D.S. – downstream river; H - habitat: P – plankton species, F – fouler; B – benthonic; Ha – halobility: hb – halophobe; i – indifferent; hl – halophile; mg – mezogalob; oh – oligogalobe; A – acidophilism: az – acidophile; i – indifferent; al – alkaliphile: Geo. – geographical distribution: k – cosmopolit, aa – arctalpine, b – boreal; C – saprobity: x – xenosaprobe, (o-x) – oligo-xenosaprobe, (x-o) – xeno-oligosaprobe, o – oligosaprobe, (o-β) – oligo- β-mezosaprobe, (β-o) – β-oligosaprobe, β – β-mezosaprobe, (β-α) – β-α-mezosaprobe, (α-β) – α-β-mezosaprobe, α – α-mezosaprobe; p – polysaprobe, «+» – species present, «-» – species not present; «?» – biographically, biologically and georgraphically understudied species)

| Species, type, from, variation | U.S. | M.S. | D.S. | Ecological and geographical characteristics | | | | |
|---|---|---|---|---|---|---|---|---|
| | | | | H | Ha | A | Geo | S |
| CYANOPHYTA (CYANOBACTERIA) | | | | | | | | |
| *Merismopedia elegans* A. Br. | - | - | + | P | i | i | k | ß |
| *M. punctata* Meyn. | - | - | + | P | i | i | k | ß |
| *M. tenuissima* Lemm. | - | + | + | P | i | i | k | ß-α |
| *Microcystis aeruginosa* Kütz. emend. Elenk. f. *aeruginosa* | + | + | + | P | i | i | k | ß |
| *M. aeruginosa* f. *flos-aquae* (Wittr.) Elenk. | + | + | + | P | i | i | k | ß |
| *M. pulverea* (Wood) Forti emend. Elenk. f. *pulverea* | + | + | + | P | i | i | k | o-ß |
| *M. pulverea* f. *incerta* (Lemm.) Elenk. | - | - | + | P | i | i | k | ß |
| *M. pulverea* f. *irregularis* (B.-Peters.) Elenk. | - | - | + | P | i | i | aa | ? |
| *M. pulverea* f. *pulchra* (Lemm.) Elenk. | - | - | + | P | i | i | aa | ? |
| *M. grevillei* f. *pulchra* (Kütz.) Elenk. | - | - | + | P | hl | ? | k | ? |
| *Aphanothece clathrata* f. *brevis* (Bachm.) Elenk. | + | + | + | P | i | i | aa | ? |
| *Gloeocapsa minuta* (Kütz.) Hollerb. ampl. | + | - | + | P | i | i | k | o |
| *Chroococcus turgidus* (Kütz.) Näg. | + | + | + | B | hl | al | k | o-ß |
| *Coelosphaerium pusillum* Goor | - | - | + | P | i | ? | k | ? |
| *Gomphosphaeria aponina* Kütz. | - | - | + | P | hl | al | k | o |
| *G. lacustris* Chod. f. *lacustris* | - | + | + | P | i | i | k | o-ß |
| *G. lacustris* f. *compacta* (Lemm.) Elenk. | - | + | + | P | i | i | b | ß |
| *Anabaena constricta* (Scaf.) Geitl. | + | + | + | P | i | i | b | p |
| *A. flos-aquae* (Lyngb.) Breb. | + | + | + | P | i | i | k | ß |
| *A. lemmermannii* P. Richt. | + | + | + | P | i | i | k | ß |
| *A. scheremetievi* Elenk. | + | + | + | P | i | al | k | o-ß |
| *A. spiroides* f. *contorta* (Kleb.) Elenk. | + | - | + | P | i | ? | k | ? |
| *A. sphaerica* Born. et Flah. | + | + | + | P | i | ? | k | o-ß |
| *Aphanizomenon flos-aquae* (L.) Ralfs | + | + | + | P | hl | i | k | ß |
| *Nostoc kihlmani* Lemm. | - | + | - | P | i | ? | k | o-ß |
| *Rivularia planctonica* Elenk. | - | - | + | ? | ? | ? | ? | ? |
| *Lyngbya kuetzingii* (Kütz.) Schmidle | - | - | + | F | hl | ? | k | o-ß |
| *L. putealis* Mont. | - | + | + | B | ? | ? | k | ? |
| *Oscillatoria limosa* Ag. | + | + | + | P | hl | al | k | ß-α |
| *O. limnetica* Lemm. | + | + | + | P | i | ? | k | o-ß |
| *Plectonema notatum* Schmidle | - | - | + | ? | ? | ? | ? | ß |

| | | | | | | | | |
|---|---|---|---|---|---|---|---|---|
| CHRYSOPHYTA | | | | | | | | |
| *Chromulina rosanoffii* (Woronin) Bütschli | - | - | + | P | hb | ? | k | o-ß |
| *Kephyrion bacilliforme* Conr. | - | - | + | ? | ? | ? | ? | ? |
| *K. boreale* Skuja | + | + | + | P | i | ? | b | o |
| *K. circumvallatum* (Schiller) Bourr. | + | + | - | ? | ? | ? | ? | o-ß |
| *K. francevii* Gus. | - | - | + | B | i | ? | ? | ? |
| *K. incostans* (Sehmid) Bourr. | + | - | - | ? | i | ? | b | ß |
| *K. laticollis* (Conr.) Bourr. | + | - | - | ? | ? | ? | ? | ? |
| *K. mosquense* Gus. | + | - | + | P | i | ? | ? | ? |
| *K. rubri-claustri* Conr. | - | - | + | B | i | ? | b | o |
| *K. spirale* (Lack.) Conr. | - | - | + | B | i | ? | ? | ß |
| *Chrysococcus rufescens* Klebs | + | + | + | P | hb | ? | k | o-ß |
| *Dinobryon bavaricum* Imh. | + | + | + | P | i | az | b | o |
| *D. cylindricum* Imh. var. *cylindricum* | + | + | + | P | i | ? | k | o-ß |
| *D. cylindricum* var. *palustre* Lemm. | + | + | + | P | i | az | b | ? |
| *D. divergens* Imh. var. *divergens* | + | + | + | P | i | i | k | ß |
| *D. divergens* var. *angulatum* (Sel.) Brunnth. | + | - | + | P | i | ? | k | ? |
| *D. elegantissimum* (Korsch.) Bourr. | + | - | + | ? | ? | ? | ? | ? |
| *D. pediforme* (Lemm.) Stein. | - | - | + | P | i | ? | b | ß |
| *D. sertularia* Ehr. | - | + | + | P | i | al | k | o |
| *D. suecicum* Lemm. var. *suecicum* | + | + | + | P | i | i | aa | o |
| *D. suecicum* var. *longispinum* Lemm. | + | + | + | ? | ? | ? | aa | ? |
| *D. sociale* Ehr. | + | - | + | P | i | az | k | ß |
| *D. spirale* Iwan. | - | + | + | ? | i | ? | k | o |
| *D. stokesii* Lemm. | + | + | - | ? | ? | ? | ? | ? |
| *Pseudokephyrion undulatissimum* Scherf. | + | - | + | P | ? | ? | ? | o |
| *Mallomonas denticulata* Matv. | - | + | - | P | hb | ? | k | ? |
| *M. caudata* Iwan. | - | + | - | P | i | az | k | ? |
| *Catenochrysis hispida* (Phillips) Perman | - | + | + | ? | ? | ? | ? | ? |
| *Synura petersenii* Korsch. | - | + | - | P | hb | az | ? | ß |
| *S. uvella* Ehr. emend. Korsch. | + | + | + | P | i | az | k | ß |
| *S. sphagnicola* Korsch. | - | - | + | P | hb | i | k | o |
| BACILLARIOPHYTA | | | | | | | | |
| *Stephanodiscus hantzschii* Grun. | + | + | + | P | i | i | k | α |
| *Cyclotella antiqua* W. Sm. | + | - | - | P | hb | az | aa | ? |
| *C. kuetzingiana* Thw. | - | - | + | P | hl | al | k | ß |
| *C. meneghiniana* Kütz. | + | + | + | P | hl | i | k | α-ß |
| *C. stelligera* (Cl. & Grun.) van Hurch | + | + | + | P | hl | i | k | ? |
| *Melosira varians* Ag. | + | + | + | P | hl | al | k | ß |
| *M. undulata* var. *normanii* Arn. | + | + | + | ? | ? | ? | ? | ? |
| *Aulacoseira distans* (Ehr.) Sim. var. *distans* | + | + | + | P | i | az | aa | x-o |
| *A. alpigena* (Crun.) Krammer =*A. distans* var. *alpigena* (Grun.) Sim. | + | + | + | P | i | i | aa | x-o |
| *A. granulata* (Ehr.) Sim. | + | + | + | P | i | al | k | ß |
| *A. italica* (Ehr.) Sim. | + | + | + | P | i | i | k | o-ß |
| *Fragilaria bicapitata* A. Mayer | - | - | + | ? | i | ? | k | o |
| *F. capucina* Desm. var. *capucina* = *Fragilaria capucina* var. *mesolepta* Rabenh. | - | - | + | P | i | i | k | ? |
| *F. capucina* var. auastrica (Raben.) Raben.= *Synedra amphicephala* var. *austriaca* Grun. | - | - | + | ? | ? | ? | ? | x |
| *F. construens* (Ehr.) Grun. | - | - | + | F | i | al | k | o-ß |
| *F. crotonensis* Kitt. | + | + | + | P | hl | al | k | o-ß |
| *F.cyclopum* (Brut.) L.-B. = *Synedra cyclopum* Brutschy | - | - | + | F | i | i | k | ? |
| *F. danica* (Kütz.) L.-B. = *Synedra ulna* var. *danica* (Kütz.) Grun. | + | + | + | F | i | al | k | o |
| *F. leptostauron* (Ehr.) Hust. | - | + | + | F | hb | i | b | ? |
| *F. virescens* var. *mesolepta* Schönf. | - | - | + | B | oh | i | ? | ? |
| *F. virescens* var. *oblongella* Grun. = F. *virescens* var. *oblongella* (Grun.) Bukht. | - | - | + | B | oh | i | ? | ? |
| *Fragilariforma virescens* (Ralfs) Will. & Round = *Fragilaria virescens* Ralfs | + | + | + | F | i | az | aa | x |
| *Pseudostaurosira brevistriata* (Crun.) Will. & Round = *Fragilaria brevistriata* Grun. | - | - | + | F | i | i | k | o |

| | | | | | | | | |
|---|---|---|---|---|---|---|---|---|
| *P. binodis* (Ehr.) Edl. = *Fragilaria construens* var. *binodis* (Ehr.) Grun. | - | + | + | F | i | al | k | o |
| *Ulnaria acus* (Kütz.) Ab. = *Synedra acus* Kütz. | + | + | + | P | i | al | k | ß |
| *Ul. biceps* (Kütz.) L.-B. = *Synedra ulna* var. *biceps* (Kütz.) Schönf. | - | - | + | F | i | al | k | ? |
| *Ul. delicatissima* var. *angustissima* (Crun.) Ab. & Sil.= *Synedra acus* var. *angustissima* Grun. | - | - | + | P | i | al | k | o |
| *Ul. ulna* (Nitsch.) Comp. var. *ulna* = *Synedra ulna* (Nitsch.) Ehr. | + | + | + | F | i | al | k | ß |
| *Ul. ulna* var. *amphirhynchus* (Ehr.) Ab. = *Synedra ulna* var. *amphirhynchus* (Ehr.) Grun. | + | + | + | F | i | al | k | ? |
| *Ul. ulna* var. *spathulifera* (Grun.) Ab. = *Synedra ulna* var. *spathulifera* Grun. | - | + | + | F | i | al | k | ? |
| *Synedra rumpens* var. *familiaris* (Kütz.) Grun. | - | + | + | B | i | i | k | ? |
| *Asterionella formosa* Hass. | + | + | + | P | i | i | k | o-ß |
| *Diatoma ehrenbergii* Kütz. = *Diatoma vulgare* var. *ehrenbergii* (Kütz.) Grun. | - | + | + | F | i | i | b | x-o |
| *D. tenuis* Kütz.= *Diatoma elongatum* (Lyngb.) Ag. | - | + | + | P | hl | i | b | o-ß |
| *D. vulgaris* Bory. | + | + | + | P | i | al | k | ß |
| *Meridion circulare* (Grev.) Ag. var. *circulare* | - | - | + | F | hb | az | aa | x-o |
| *M. circulare* var. *constrictum* (Ralfs) V. H. | + | + | + | F | hb | az | aa | ? |
| *Tabellaria fenestrata* (Lyngb.) Kütz. | + | + | + | P | hb | az | k | o-ß |
| *T. flocculosa* (Roth) Kütz. | + | + | + | F | hb | az | aa | o-x |
| *Tetracyclus lacustris* Ralfs var. *lacustris* | + | + | + | F | i | az | aa | ? |
| *T. lacustris* var. *capitata* Hust. | - | + | + | F | hb | az | aa | ? |
| *Adlafia minuscula* (Grun.) L.-B. = *Navicula minuscula* Grun. | - | + | + | B | i | al | k | ß |
| *Caloneis molaris* (Grun.) Krammer = *Pinnularia molaris* Grun. | - | + | - | B | i | i | aa | ? |
| *Cal. silicula* (Ehr.) Cl. | + | + | + | B | i | al | k | o-ß |
| *Cal. undulata* Greg. Krammer = *P. undulata* var. *subundulata* Grun. | + | - | - | ? | ? | ? | ? | ? |
| *Cavinula lacustris* (Greg.) Mann & Stick. = *Navicula lacustris* Greg. | + | + | + | B | i | i | b | o |
| *Cav. pusio* (Cl.) L.-B. = *Navicula pusio* Cl. | - | + | + | B | i | i | b | ? |
| *Cosmineis pusilla* (Smith) Mann & Stick. = *Navicula pusilla* W. Sm. | - | + | + | B | i | i | k | ? |
| *Craticula ambigua* (Ehr.) Mann = *Navicula bicapitellata* Hust. | - | + | + | B | i | ? | k | ? |
| *Cr. cuspidata* (Kütz.) Mann & Bound var. *cuspidata* f. *cuspidata* = *Navicula cuspidata* Kütz. | + | + | + | B | i | i | k | α |
| *Cr. cuspidata* f. *primigena* Dipp. = *Navicula cuspidata* f. *primigena* Dipp. | + | + | + | B | i | i | b | ? |
| *Decussata placenta* (Ehr.) L.-B. = *Navicula placenta* Ehr. | + | + | + | B | hb | az | k | ? |
| *Diploneis ovalis* (Hilse) Cl. | - | + | + | B | i | i | b | ? |
| *Frustulia saxonica* Raben. = *Frustulia rhomboides* (Ehr.) D. T. | - | - | + | B | hb | az | aa | o-ß |
| *F. vulgaris* Thw. D.T. = *Frustulia rhomboides* f. *vulgaris* Thw. | + | + | + | B | hb | i | b | o |
| *Gyrosigma acuminatum* (Kütz.) Raben. var. *acuminatum* | + | + | + | B | i | al | b | ß |
| *G. acuminatum* var. *gallicum* Grun. | - | + | + | B | hl | ? | k | ? |
| *Hippodonta capitata* (Ehr.) L.-B., Metzeltin & Witkowski = *Navicula hungarica* var. *capitata* Cl. | + | + | + | B | hl | al | b | ß |
| *H. costulata* (Grun.) L.-B., Metzeltin & Witkowski = *Navicula costulata* Grun. | - | + | + | B | hl | al | b | ? |
| *Mayamaea atomus* (Kütz.) L.-B.= *Navicula atomus* (Näg.) Grun. | - | - | + | B | hl | al | k | ß-α |
| *Navicula cryptocephala* Kütz. var. *cryptocephala* | + | + | + | B | i | al | k | α |
| *N. cryptocephala* var. *lata* Poret. et Anissim. | - | - | + | B | ? | ? | ? | ? |
| *N. gastrum* Ehr. | - | + | + | B | i | i | k | ß |
| *N. gibbula* Cl. | - | + | + | B | i | ? | aa | ? |
| *N. gracilis* Ehr. | + | + | + | B | i | i | b | o-ß |
| *N. oblonga* (Kütz.) Kütz. = *Navicula oblonga* Kütz. | + | - | + | ? | ? | ? | ? | ? |
| *N. platystoma* Ehr. | - | - | + | B | i | ? | k | ? |
| *N. radiosa* Kütz. | + | + | + | B | i | i | k | o-ß |
| *N. reinhardtii* (Grun.) Cl. | + | + | + | B | i | al | b | o-ß |
| *N. viridula* Kütz. | + | + | + | B | hl | al | k | α |

| | | | | | | | | |
|---|---|---|---|---|---|---|---|---|
| *Neidium affine* (Ehr.) Pfitzer f. *affine* = *Neidium affine* (Ehr.) Cl. f. *affine* | + | + | + | B | i | al | b | ? |
| *N. affine* f. *capitatum* Skv. et Meyer | - | - | + | ? | ? | ? | ? | ? |
| *N. iridis* (Ehr.) Cl. = *N. iridis* var. *amphigomphus* (Ehr.) V. H. | + | + | + | B | hb | i | b | ? |
| *Pinnularia acrosphaeria* Breb. | + | - | + | B | i | al | k | ? |
| *P. angusta* (Cl.) Krammer = *Pinnularia mesolepta* f. *angustata* Cl. | - | + | - | B | i | i | b | ? |
| *P. appendiculata* (Ag.) Cl. | + | + | + | B | hl | i | b | ? |
| *P. biceps* Greg. = *Pinnularia interrupta* W. Sm. f. *interrupta* | - | + | + | B | i | i | b | o-ß |
| *P. borealis* Ehr. | + | + | + | B | i | i | aa | ? |
| *P. brauniana* (Grun.) Mills = *Pinnularia braunii* var. *amphicephala* (A. Mayer) Hust. | - | - | + | B | hl | i | aa | ? |
| *P. brebissonii* (Kütz.) Raben. = *Pinnularia microstauron* var. *brebissonii* (Kütz.) Hust. | + | + | + | B | i | i | b | ? |
| *P. fasciata* (Lagerst.) Hust. | + | + | + | ? | ? | ? | aa | ? |
| *P. gibba* Ehr. var. *gibba* f. *gibba* | + | + | + | B | i | i | b | x-o |
| *P. gibba* f. *subundulata* A. Mayer | + | - | + | B | i | i | b | ? |
| *P. gracillima* Greg. | - | + | + | B | hl | i | k | x |
| *P. mesolepta* (Ehr.) W. Sm. | - | + | - | B | i | al | b | o |
| *P. microstauron* (Ehr.) Cl. var. *microstauron* = *Pinnularia gibba* var. *parva* (Ehr.) Grun.; *P. microstauron* (Ehr). Cl. | + | - | + | B | i | i | b | ? |
| *P. microstauron* var. *ambigua* Meist. | + | - | + | ? | ? | ? | ? | ? |
| *P. neomajor* Krammer = *Pinnularia major* (Kütz.) Cl. var. *major*; *Pinnularia major* var. *lacustris* Meist. | - | + | + | B | i | i | b | ß |
| *P. interruptiformis* Krammer = *Pinnularia interrupta* f. *minor* Boye P. | - | + | - | B | i | i | b | ? |
| *P. stauroptera* (Grun.) Raben. | - | + | + | B | ? | az | k | o |
| *P subcapitata* Greg. = *P. subcapitata* var. *hilseana* (Janisch.) O. Müll. | - | + | + | B | hl | az | k | x-o |
| *P. subgibba* Krammer = *Pinnularia gibba* var. *linearis* Hust. | + | - | + | B | i | az | b | ? |
| *P. sublinearis* Grun. | - | + | + | B | i | ? | b | ? |
| *P. subsolaris* (Grun.) Cl. | + | + | + | B | oh | i | aa | ? |
| *P. viridis* (Nitzsch.) Ehr. | + | + | + | B | i | i | b | o-ß |
| *Placoneis dicephala* (Ehr.) W. Sm. = *Navicula dicephala* (Ehr.) W. Sm. | + | - | + | B | i | i | k | o-ß |
| *Pl. elliptica* Ohtsuka = *Navicula exigua* var. *elliptica* Hust. | - | + | + | B | i | ? | k | ? |
| *Sellaphora pupula* (Kütz.) Meresch. = *Navicula pupula* Kütz. | - | + | + | B | hl | i | k | ß |
| *Sel. parapupula* L.-B. in L.-B.et Metz. = *Navicula pupula* var. *capitata* Hust. | - | + | + | B | hl | i | k | ? |
| *Sel. rectangularis* (Greg.) L.-B. Meresch. = *Navicula pupula* var. *rectangularis* (Greg.) Grun. | + | + | + | B | hl | i | k | ? |
| *Stauroneis acuta* W. Sm. | + | - | + | B | i | al | k | o |
| *S. anceps* Ehr. f. *anceps* | + | + | + | B | i | i | k | ß |
| *S. gracilior* (Reich.) Metz., L.-B.= *Stauroneis anceps* f. *gracilis* (Ehr.) Cl. | + | + | + | B | i | i | k | ? |
| *S. parvula* Grun. | - | + | + | B | i | i | aa | ? |
| *S. phoenicenteron* (Nitzsch.) Ehr. = *Stauroneis phoenicenteron* Ehr. | + | + | + | B | i | i | b | ß |
| *Achnanthes inflata* (Kütz.) Grun. | + | - | - | ? | ? | ? | ? | ? |
| *Cocconeis placentula* Ehr. var. *placentula* | + | + | + | F | i | al | k | o-ß |
| *C. placentula* var. *euglypta* (Ehr.) Cl. | - | - | + | F | i | az | b | ? |
| *Planothidium ellipticum* (Cl.) Edl. = *Achnanthes lanceolata* var. *elliptica* Cl. | - | + | + | F | i | i | aa | ? |
| *Pl. lanceolatum* (Breb.) Round et Bukht. f. *lanceolatum* = *Achnanthes lanceolata* (Breb.) Grun.; *A. lanceolata* f. *ventricosa* Hust. | + | + | + | F | i | al | k | o-ß |
| *Pl. lanceolatum* f. *capitata* (O. Müll.) L.-B. = *Achnanthes lanceolata* f.*capitata* (O. Müll.) Hust. | - | + | - | F | ? | ? | ? | ? |
| *Pl. rostratum* (Øest.) Round et Bukht. = *Achnanthes lanceolata* f. *rostrata* (Østr.) Hust. | + | - | + | F | i | i | k | ? |
| *Eunotia arcus* Ehr. = *Eunotia arcus* var. *bidens* Grun. | + | + | + | F | i | al | k | ? |
| *E. bigibba* Kütz. = *E. bigibba* var. *pumila* Grun. | - | - | + | F | i | az | aa | ? |

| | | | | | | | | |
|---|---|---|---|---|---|---|---|---|
| *E. diodon* Ehr. | + | + | + | F | ? | az | aa | ? |
| *E. elegans* Østr. | + | + | + | F | ? | ? | ? | ? |
| *Eunotia gracilis* (Ehr.) Rabenh.= *Eunotia exigua* (Breb.) Rabenh. var. *exigua* | + | - | + | F | hb | az | aa | ß |
| *E. variondulata* Nörpel-Schempp & L.-B. = *Eunotia exigua* var. *tridentula* Østr. | + | - | - | F | hb | az | k | ß |
| *E. faba* var. *densestriata* Østr. | - | + | + | F | hb | ? | b | ? |
| *E. groenlandica* (Grun.) Nörpel-Schempp & L.-B. = *Eunotia fallax* var. *gracillima* Krasske | - | - | + | F | hb | i | k | ? |
| *E. formica* Ehr. | + | + | + | F | i | az | k | o |
| *E. bilunaris* (Ehr.) Schaar. = *Eunotia lunaris* (Ehr.) Grun.; *E. lunaris* var. *capitata* Grun. | + | + | + | F | i | az | k | o |
| *E. mucophila* (L.-B. & Nörpel) L.-B. = *Eunotia lunaris* var. *subarcuata* (Näg.) Grun. | + | + | + | F | i | az | k | ? |
| *E. monodon* (Fontell) Cl., L.-B. Metz. = *Eunotia monodon* Ehr.; *E. monodon* var. *major* (W. Sm.) Hust. | + | + | + | F | hb | az | k | o |
| *E. jemtlandica* (Fontell) Cl., L.-B. Metz. = *Eunotia monodon* var. *bidens* (Greg.) W. Sm. | + | + | + | F | hb | az | k | ? |
| *E. parallela* Ehr. | + | + | + | F | i | i | aa | o |
| *E. pectinalis* (Dillw.? Kütz.) Rabenh. var. *pectinalis* | + | + | + | F | i | i | k | o-ß |
| *E. undulata* W. Smith; Krammer; L.-B. = *Eunotia pectinalis* var. *ventralis* (Ehr.) Hust.; *E. pectinalis* var. *undulata* Ralfs | + | + | + | F | i | i | k | ? |
| *E. paratridentula* L.-B.& Kulikov. = *Eunotia polydentula* Brun. | - | + | - | F | hb | az | k | x-o |
| *E. septena* Ehr. = *Eunotia polyglyphis* Grun. | - | + | + | F | hb | az | aa | ? |
| *E. praerupta* Ehr. = *Eunotia praerupta* var. *inflata* Grun. | + | + | + | F | hb | az | k | ? |
| *E. bidens* Ehr.; L.-B.; Metz. = *Eunotia praerupta* var. *bidens* (W. Sm.) Grun | + | + | + | F | hb | az | k | ? |
| *E. curtagrunowii* Nörpel – *Sehempp & L.-B.* = *Eunotia praerupta* var. *muscicola* Boye P. | + | + | + | F | hb | az | aa | ? |
| *E. serra* Ehr. = *Eunotia robusta* Ralfs | + | + | + | F | hb | az | aa | o |
| *E. neosibirica L.-B.; Kulikov.; & Witrow.* = *Eunotia sibirica* Cl. | + | + | + | F | i | ? | b | ? |
| *E. septentrionalis* Østr. | - | + | + | F | i | az | aa | ? |
| *E. sudetica* O. Müll. = *Eunotia sudetica* var. *bidens* Hust. | + | + | + | F | ? | ? | b | - |
| *E. tenella* (Grun.) Hust. | + | + | + | F | hb | az | aa | o |
| *E. veneris* (Kütz.) D.T. | + | + | + | F | hb | az | aa | o |
| *Rhoicosphenia curvata* (Kütz.) Grun. | - | - | + | F | hl | al | k | ß |
| *Amphora ovalis* Kütz. var. *ovalis* | + | + | + | B | i | al | k | o-ß |
| *A. ovalis* var. *gracilis* Ehr. | + | - | - | B | i | al | k | ß-o |
| *Cymbella aspera* (Ehr.) Cl. | - | - | + | F | i | i | aa | ß |
| *C. lanceolata* (Ehr.) V. H. | + | + | - | B | i | al | b | ß |
| *C. naviculiformis* Auersw. | + | + | + | F | i | al | b | ß |
| *C. parva* (W. Sm.) Cl. | - | + | - | F | i | i | b | ? |
| *Cymbopleura amphicephala* (Naegeli) Krammer = *Cymbella amphicephala* Näg. | - | - | + | F | i | i | b | ? |
| *Cymbop. angustata* (W. Sm.) Krammer = *Cymbella angustata* (W. Sm.) Cl. | - | - | + | ? | ? | ? | ? | ? |
| *Cymbop. arctica* (Lager.) Scmidt; Krammer; L.-B. = *Cymbella cistula* var. *arctica* Lagerst. | - | + | + | B | i | al | aa | ? |
| *Cymbop. cuspidata* (Kütz.) Krammer = *Cymbella cuspidata* Kütz. | + | + | + | B | i | i | k | ? |
| *Eucyonema minutum* (Hilse) Mann = *Cymbella ventricosa* Kütz. | + | + | + | F | i | i | k | ß |
| *Gomphonema acuminatum* Ehr. var. *acuminatum* | + | + | + | F | i | az | b | ß |
| *G. acuminatum* var. *coronatum* (Ehr.) W. Sm. | + | + | + | F | i | i | b | ß |
| *G. angustum* Kütz. = *Gomphonema intricatum* Kütz. | + | - | - | F | i | i | b | o |
| *G. olivaceum* (Lyngb.) Kütz. var. *olivaceum* | - | + | + | F | i | i | b | ß |
| *G. olivaceum* var. *calcareum* Cl. | - | - | + | F | i | i | b | ß |
| *G. subclavatum* (Kütz.) Raben. = *Gomphonema longiceps* var. *subclavatum* Grun. | + | - | - | F | i | i | b | ? |
| *Didymosphenia geminata* (Lyngb.) M. Schmidt | + | + | + | F | i | i | a | x |
| *Rhopalodia gibba* (Ehr.) O. Müll. var. *gibba* | + | + | + | F | i | i | k | o |
| *R. gibba* var. *ventricosa* (Ehr.) Grun. | + | - | - | F | hl | al | k | ? |
| *Nitzschia acuta* Hantzsch | - | + | + | B | i | i | b | ß |

| | | | | | | | | |
|---|---|---|---|---|---|---|---|---|
| *N. palea* ( Kütz.) W. Sm. | + | + | + | B | i | i | b | α |
| *N. forticola* Grun. | + | - | - | B | i | i | b | o-ß |
| *N. gracilis* Hantzsch | + | - | + | B | i | i | b | ß-o |
| *N. intermedia* Hantzsch | - | - | + | ? | ? | ? | ? | ? |
| *N. sublinearis* Hust. | - | - | + | B | i | i | b | o-ß |
| *Hantzschia amphioxys* (Ehr.) Grun. | + | + | + | B | i | al | k | α |
| *Surirella angustata* Kütz. var. *angustata* | + | + | + | B | i | i | b | ß |
| *S. angustata* var. *constricta* Hust. | + | - | + | ? | ? | ? | ? | ? |
| *S. biseriata* Breb. var. *biseriata* | + | + | + | B | i | az | k | ß |
| *S. biseriata* var. *constricta* Grun. | - | + | - | B | i | i | b | ? |
| *S. capronii* Breb. | - | - | + | B | i | i | k | ß |
| *S. gracilis* (W. Sm.) Grun. | - | + | + | B | i | i | b | ? |
| *S. linearis* W. Sm. | + | + | + | B | i | i | b | ß |
| *S. grunowii* Kul.; L.-B. & Witk.= *Surirella linearis* var. *constricta* (Ehr.) Grun. | + | - | + | B | i | i | b | ß |
| *S. ovalis* Breb. | - | + | + | B | mg | i | b | ? |
| *S. robusta* Ehr. | + | + | + | B | i | i | b | ß |
| *S. splendida* (Ehr.) Kütz. = *Surirella robusta* var. *splendida* Ehr. | - | - | + | B | i | i | b | ß |
| *S. tenera* Greg. | - | + | + | B | i | i | b | ß |
| *Cymatopleura solea* (Breb.) W. Sm. var. solea | + | + | + | B | i | al | k | ß-α |
| *C. solea* var. *vulgaris* Meist. | - | - | + | B | i | i | b | ? |
| *Stenopterobia intermedia* var. *capitata* Font. | + | + | + | B | ? | ? | ? | ? |
| EUGLENOPHYTA | | | | | | | | |
| *Euglena acus* Ehr. | + | + | + | P | i | i | k | ß |
| *E. oxyuris* Schmarda | - | + | - | ? | ? | ? | ? | o-ß |
| *Phacus acuminatus* var. *acuticauda* (Roll) Pochm. | - | + | + | ? | i | ? | k | ? |
| *P. agilis* Skuja | - | - | + | ? | i | ? | k | ß |
| *P. caudatus* Hübner var. *caudatus* | - | + | + | ? | i | i | k | ß |
| *P. caudatus* var. *tenuis* Swir. | - | - | + | ? | ? | ? | ? | ? |
| *P. curvicauda* Swir. | - | - | + | P | i | i | k | ß |
| *P. longicauda* (Ehr.) Duj. | - | - | + | P | i | i | k | ß-α |
| *P. monilatus* var. *suecicus* Lemm. | - | - | + | ? | hb | al | b | ? |
| *P. orbicularis* f. *communis* Popova | - | + | + | ? | ? | ? | ? | ? |
| *P. pleuronectes* (Ehr.) Duj. | - | + | + | ? | i | i | k | ß |
| *Trachelomonas caudata* (Ehr.) Stein | - | - | + | ? | hb | ? | k | ß |
| *T. fusiformis* Stokes | - | - | + | P | i | ? | k | ? |
| *T. hispida* (Perty) Stein emend. Defl. | + | + | + | P | i | i | k | ß |
| *T. intermedia* Dang. | - | - | + | P | i | i | k | ß |
| *T. lacustris* Drez. emend. Balech | - | - | + | P | hb | i | k | o-ß |
| *T. planctonica* Swir. f. *planctonica* | + | + | + | P | i | al | k | ß-o |
| *T. planctonica* f. *oblonga* (Drez.) Popova | + | + | + | P | i | ? | k | ß |
| *T. pavlovskoënsis* (Poljansk.) Popova | - | - | + | ? | ? | ? | ? | ß-α |
| *T. volvocina* Ehr. | + | + | + | P | hl | i | k | ß |
| *T. volvocinopsis* Swir. | - | - | + | P | i | i | k | ß |
| *Strombomonas acuminata* (Schmarda) Defl. | + | + | + | P | hl | i | k | ß |
| *S. fluviatilis* (Lemm.) Defl. | - | - | + | P | i | i | k | ß |
| *S. shauinslandii* (Lemm.) Defl. | - | - | + | P | i | ? | k | ß |
| *Lepocinclis fusiformis* (Carter) Lemm. | - | - | + | P | hl | ? | k | ß |
| *L. ovum* (Ehr.) Mink. | + | - | + | ? | i | i | k | α-ß |
| XANTHOPYTA | | | | | | | | |
| *Isthmochloron trispinatum* (W. et G. S. West) Skuja | - | + | + | ? | ? | ? | ? | ? |
| *Tetraëdriella spinigera* Skuja | + | + | - | ? | ? | ? | ? | ? |
| *Pseudostaurastrum hastatum* (Reinsch) Chod. | + | - | - | ? | ? | ? | ? | ? |
| *Centritractus belonophorus* Lemm. | + | - | - | ? | ? | i | k | o-ß |
| *Ophiocytium capitatum* Wolle | - | - | + | P | i | ? | k | o |
| *O. lagerheimii* Lemm. | + | + | + | ? | ? | i | k | ? |
| *Characiopsis naegelli* (A. Br.) Lemm. | - | - | + | ? | ? | ? | ? | ? |
| DINOPHYTA | | | | | | | | |
| *Ceratium hirundinella* (O.F.M.) Bergh var. *hirundinella* | + | - | + | P | hb | az | k | o |
| *C. hirundinella* f.*gracile* Bachm. | + | + | + | P | i | ? | k | o |
| *Katodinium vorticella* Stein | - | - | + | ? | ? | ? | ? | ß-α |
| CHLOROPHYTA | | | | | | | | |

| | | | | | | | | |
|---|---|---|---|---|---|---|---|---|
| *Eudorina elegans* Ehr. | + | + | + | P | i | i | k | ß |
| *Pandorina charkowiensis* Korsch. | - | - | + | P | i | i | ? | ß |
| *P. morum* (Müll.) Bory | + | + | + | P | i | i | k | ß |
| *Sphaerocystis planctonica* (Korsch.) Bourr. | + | - | + | P | i | ? | k | ? |
| *Heleochloris pallida* Korsch. | - | + | + | ? | ? | ? | ? | ? |
| *Mychonastes jurisii* Hind.= *Dactylosphaerium jurisii* Hind. | + | + | + | P | hb | az | k | ? |
| *Ankyra judayi* (G. M. Smith) Fott | - | - | + | P | i | ? | k | ß |
| *Desmatractum indutum* (Geitl.) Rasch. | + | + | + | ? | ? | ? | ? | ? |
| *Hydrodictyon reticulatum* (L.) Lagerh. | + | - | - | ? | ? | ? | ? | ? |
| *Pediastrum angulosum* (Ehr.) Menegh. | + | + | + | P | i | i | k | o |
| *P. duplex* Meyen var. *duplex* = *Pediastrum biradiatum* Meyen | + | + | + | P | i | i | k | ß |
| *P. duplex* var. *cornutum* Racib. = *Pediastrum boryanum* var. *cornutum* (Racib.) Sulek | + | + | - | P | ? | i | k | ? |
| *P. duplex* var. *subgranulatum* Rac. | - | - | + | ? | ? | ? | ? | ? |
| *P. boryanum* (Turp.) Menegh. var. *boryanum* | + | + | + | P | i | i | k | ß |
| *P. boryanum* var. *longicorne* Reinsch | + | + | + | P | i | i | k | ? |
| *P. braunii* Wartm. | + | + | + | ? | ? | ? | ? | ? |
| *P. tetras* (Ehr.) Ralfs | + | + | + | P | i | i | k | ß |
| *Pseudopediastrum kawraiskyi* Schmidle = *Pediastrum kawraiskyi* Schmidle | + | + | + | P | i | i | b | o-ß |
| *Lacunastrum gracillimum* (W. & G. S. West) Manus = *Pediastrum duplex* var. *gracillimum* W. & G. S. West | + | + | + | P | i | ? | k | ß |
| *Monactinus simplex* (Meyen) Corda = *P. simplex* Meyen | - | + | + | P | i | ? | k | o-ß |
| *Golenkinia radiata* Chod. | + | + | + | P | i | i | k | ß |
| *Golenkiniopsis longispina* (Korsch.) Korsch. | + | - | - | P | i | ? | k | ? |
| *Micractinium bornhemiense* (Corn.) Korsch. | - | - | + | P | i | ? | k | ß |
| *M. pusillum* Fres. | - | - | + | P | ? | ? | k | ß |
| *M. quadrisetum* (Lemm.) G. M. Smith | - | - | + | P | ? | ? | k | ? |
| *Botryosphaera sudetica* (Lemm.) Chod. | + | + | + | B | hb | az | b | ? |
| *Mucidosphaerium pulchellum* (Wood) Bock, Proschold & Krienitz = *Dictyosphaerium pulchellum* Wood | + | + | + | P | i | i | k | ß-o |
| *Dictyosphaerium tetrachotomum* Printz | - | + | - | P | i | i | k | ? |
| *Radiococcus polycoccus* (Hind.)= *Sphaerocystis polycocca; Coenococcus polycoccus* (Korsch.) Hind.Korsch. | - | - | + | P | i | ? | k | ? |
| *Coenococcus planctonicus* Korsch. | + | + | + | P | i | i | k | ? |
| *Tetraëdron caudatum* (Corda) Hansgirg | + | - | - | P | i | i | k | ß |
| *T. minimum* (A. Br.) Hansg. | + | + | + | P | i | ? | k | ß |
| *T. triangulare* Korsch. | - | - | + | P | i | ? | k | ß |
| *Chlorotetraedron* incus (Teil.) Komárek & Kovácik = *Tetraëdron incus* (Teil.) G. M. Smith | + | + | + | P | i | i | k | ß |
| *Franceia ovalis* Lemm. = *Franceia tenuspina* Korsh. | - | + | - | P | i | ? | k | ? |
| *Lagerheimia ciliata* (Lagerh.) Chod. | - | - | + | P | i | i | k | ß |
| *L. longiseta* (Lemm.) = *Lagerheimia longiseta* (Lemm.) Wille; *Lagerheimia citriformis* (Snow) Collins | + | + | + | P | i | ? | k | ß |
| *L. genevensis* (Chod.) Chod. | - | + | + | P | i | ? | k | ß |
| *L. subsalsa* Lemm. | + | + | + | P | i | ? | k | ? |
| *Oocystis borgei* Snow | - | + | + | P | i | i | k | ß-o |
| *O. solitaria* Wittr. | - | - | + | P | i | i | k | ß-o |
| *Monoraphidium arcuatum* (Korsch.) = *Monoraphidium arcuatum* (Korsch.) Hindak | - | + | + | P | i | i | k | ß |
| *M. contortum* (Thur.) Kom.-Legn. | - | + | + | P | i | ? | k | ß |
| *M. griffithii* (Berk.) Kom.-Legn. | - | + | + | P | i | ? | k | ß |
| *M. irregulare* (G. M. Smith) Kom.-Legn. | + | + | + | P | i | i | k | ? |
| *Kirchneriella aperta* Teil. | + | + | + | P | i | i | k | ? |
| *K. lunaris* (Kirchn.) Möb. | - | - | + | P | i | ? | k | ß |
| *K. obesa* (W. West) Schmidle | + | + | + | P | i | i | k | ß |
| *Raphidocelis contorta* (Schmidle) Marvan, Komarek, Comas | + | - | - | P | ? | ? | k | ß |
| *R. danubiana* (Hind.) Marvan, Komarek, Comas | + | + | + | P | ? | ? | ? | ? |
| *Quadrigula korschikoffii* Kom. | + | + | + | ? | ? | ? | ? | ? |
| *Ankistrodesmus falcatus* (Corda) Ralfs | - | + | + | P | hb | az | b | ß-α |
| *A. fusiformis* Corda ex Korsch. | + | + | + | P | i | ? | k | ß |
| *A. spiralis* (Turp.) Lemm. | + | + | + | P | i | ? | k | ß |

IOP Conf. Series: Earth and Environmental Science **138** (2018) 012017 doi:10.1088/1755-1315/138/1/012017

| | | | | | | | | |
|---|---|---|---|---|---|---|---|---|
| *Selenastrum gracile* Reinsch | + | + | + | P | i | ? | k | ß |
| *Hyaloraphidium contortum* Rasch. et Korsch. var. *contortum* | + | + | + | P | i | ? | k | ? |
| *H. contortum* var. *tenuissimum* Korsch. | - | + | - | ? | i | ? | k | ? |
| *Coelastrum astroideum* De-Not. | + | - | + | P | i | i | k | ß |
| *C. pulchrum* Schmidle | - | - | + | P | ? | ? | ? | ? |
| *C. microporum* Näg. | + | + | + | P | i | i | k | ß |
| *C. sphaericum* Näg. | + | + | + | P | i | i | k | o-ß |
| *Actinastrum hantzschii* Lagerh. var. *hantzschii.* | + | + | + | P | i | i | k | ß |
| *A. hantzschii* Lagerh. var. *subtille* Wolosz. | - | + | + | P | i | ? | k | ? |
| *Crucigenia fenestrata* (Schmidle) Schmidle | + | + | + | P | i | ? | k | ß |
| *C. quadrata* Morr. | + | + | + | P | i | i | k | ? |
| *C. tetrapedia* (Kirchn.) W. et G. S. West | + | + | + | P | i | i | k | o-ß |
| *Crucigeniella apiculata* (Lemm.) Schmidle | + | + | + | P | i | ? | k | ß |
| *Cr. irregularis* (Wille) Tsarenko & John = *Willea irregularis* (Wille) Schmidle | + | - | + | P | i | ? | k | ? |
| *Tetrastrum elegans* Playf. | - | + | + | P | i | ? | k | o-ß |
| *T. komarekii* Hind. | + | + | + | P | i | i | k | ß |
| *T. staurogeniaeforme* (Schröd.) Lemm. | - | + | + | P | i | ? | k | ß |
| *T. triacanthum* Korsch. | + | + | + | P | i | ? | k | ß |
| *T. triangulare* (Chod.) Kom. | + | + | + | P | i | ? | k | ß |
| *Pseudodidymocystis lineata* (Korsh.) = *Didymocystis lineata* Korsch. | - | - | + | P | i | ? | k | ? |
| *Ps. planctonica* (Korsch.) = *Didymocystis planctonica* Korsch. | - | - | + | P | i | i | k | ß |
| *Acutodesmus acuminatus* (Lagerch.) Tsarenko Balwois = *Scenedesmus acuminatus* (Lagerch.) Chod.; *Scenedesmus acuminatus* var. *elongatus* (G. M. Smith) Deduss.; *Scenedesmus acuminatus* var. *tortuotus* Skuja | + | + | + | P | i | i | k | ß |
| *Ac. obliquus* (Tupr.) Heg. & Hanag.= *Scenedesmus obliquus* (Tupr.) Kütz. | + | + | + | P | i | ? | k | ß |
| *Enallax acutiformis* (Schröd.) Hindák. According to Hegewald = *Scenedesmus acutiformis* Schröd. | + | + | + | P | i | ? | k | ß |
| *Scenedesmus apiculatus* (W. et G. S. West) Chod. | + | + | + | P | i | ? | k | ? |
| *S. armatus* (Chod.) Chod. = *Scenedesmus helveticus* Chod. | - | - | + | P | i | ? | k | ß |
| *S. arcuatus* (Lemm.) Lemm. | + | + | + | P | i | ? | k | ß |
| *S. serratus* (Corda) Rohlin = *Scenedesmus denticulatus* var. *linearis* Hansg. | + | + | + | P | i | i | k | ß |
| *S. ellipticus* Corda | + | + | + | P | i | ? | ? | ? |
| *S. acuminatus* (Lagerh.) Chodat = *Scenedesmus falcatus* Chod. | + | + | + | ? | ? | ? | ? | ? |
| *S. subspicatus* Chod. = *Scenedesmus gutwinskii* Chod. | + | + | + | P | i | ? | k | o-ß |
| *S. magnus* Meyen var. *magnus* = *Scenedesmus quadricauda* var. *setosus* Kirchn. | + | + | + | P | i | i | k | o-ß |
| *S. magnus* var. *naegelii* (Breb.) Tzar. | + | + | + | ? | ? | ? | ? | ? |
| *S. microspina* Chod. | + | + | + | ? | i | ? | ? | ? |
| *S. obtusus* Meyen | - | - | + | P | i | i | k | ß |
| *S. semipulcher* Hortob. | + | - | + | P | i | ? | k | ß |
| *S. abundans* .(Kirch.) Chod. = *Scenedesmus sempervirens* Chod. | + | + | + | P | i | i | k | ? |
| *S. quadricauda* (Turp.) Breb. | + | + | + | P | i | i | k | ß |
| *Scenedesmus obliguus* (Turp.) = *Scenedesmus acutus* Meyen | + | + | + | P | i | ? | k | ß |
| *S. caudato-aculeolatus* Chod. | - | + | + | P | i | i | k | ? |
| *S. circumfusus* Hortob. var. *bicaudatus* | - | - | + | ? | ? | ? | ? | ? |
| *Desmodesmus armatus* (Chod.) Hegew var. *armatus* = *Scenedesmus armatus* Chod. var. *armatus* | + | + | + | P | i | ? | k | ß |
| *Des. armatus* var. *bicaudatus* (Ugl.) Egew. = *Scenedesmus armatus* var. *bicaudatus* (Gugl.) Chod. | - | - | + | ? | ? | ? | ? | ? |
| *Des.brasiliensis* (Bohl.) Egew. *Scenedesmus brasiliensis* Bohl. | + | + | + | ? | ? | ? | k | ß |
| *Des. denticulatus* (Lagerh.), Friedl et Egew. = *Scenedesmus denticulatus* Lagerh. var. *denticulatus* | + | + | + | P | i | i | k | ß |
| *Des.hystrix* (Lagerh.) Hegew. = *Scenedesmus hystrix* Lagerh. | + | - | + | P | ? | al | ? | ß |
| *Des. intermedius* (Chod.) Hegew. var *intermedius* = *Scenedesmus intermedius* Chod. var. *intermedius* | + | + | + | P | i | i | k | ? |
| *Des. intermedius* var *acutispinus* (Roll) Hegew. = *Scenedesmus intermedius* var. *bicaudatus* Hortob. | + | + | + | ? | ? | ? | ? | ? |

| | | | | | | | | |
|---|---|---|---|---|---|---|---|---|
| *Des. lefevrei* (Defl.) Friedl *et* Hegew. = *Scenedesmus lefevrii* Defl. | + | + | + | P | i | ? | k | ß |
| *Des. opoliensis* (Richt.) Hegew. var *opoliensis* = *Scenedesmus opoliensis* P. Richt var. *opoliensis* | + | - | - | P | ? | i | k | ß |
| *Des. opoliensis* var. *carinatus* (Lemm.) Hegew. = *Scenedesmus opoliensis* var. *carinatus* Lemm. | - | - | + | P | ? | ? | ? | ? |
| *Des. spinosus* (Chod.) Hegew. = *Scenedesmus spinosus* Chod. | + | + | + | P | i | ? | k | ? |
| *Ulothrix teneririma* Kütz. = *Ulothrix variabilis* Kütz. | + | - | + | B | i | ? | k | ? |
| *Closterium abruptum* f. *brevius* W. et G. West | + | - | - | ? | ? | ? | ? | ? |
| *C. acutum* (Lyngb.) Breb. | + | + | + | P | i | i | k | ß |
| *C. acerosum* (Schr.) Ehr. f. *acerosum* | - | + | + | B | i | i | k | ß |
| *C. acerosum* f. *elongatum* (Breb.) Kossinsk. | - | + | + | P | i | ? | k | ? |
| *C. aciculare* Tuffen West | + | + | + | P | i | i | k | ß |
| *C. angustatum* Kütz. | - | + | - | ? | ? | ? | ? | ? |
| *C. dianae* Ehr. | - | + | - | P | i | ? | k | o |
| *C. ehrenbergii* Menegh. | - | + | + | ? | hb | i | k | ß |
| *C. ehrenbergii* var. *percrassum* (Borge) Grönbl. | - | + | - | ? | ? | ? | ? | ? |
| *C. gracile* Breb. f. *Gracile* | + | + | + | P | hb | ? | k | o |
| *C. gracile* f. *elongatum* (W. et G. West.) Kossinsk. | - | - | + | P | hb | ? | k | ? |
| *C. incurvum* Breb. | - | - | + | ? | ? | ? | ? | ? |
| *C. jenneri* Ralfs | + | + | + | ? | ? | ? | ? | ? |
| *C. lanceolatum* Kütz. | - | + | - | P | hb | ? | k | ? |
| *C. leibleinii* Kütz. | + | - | + | P | i | ? | k | α |
| *C. littorale* Gay | + | + | - | P | ? | ? | ? | ? |
| *C. moniliferum* (Bory) Ehr. var. *moniliferum* | + | + | + | B | i | ? | k | ß |
| *C. moniliferum* var. *concavum* Klebs | - | + | + | P | i | al | k | ? |
| *C. parvulum* Näg. f. *parvulum* | - | + | + | ? | i | i | k | ß |
| *C. parvulum* f. *majus* W. West | - | + | + | P | i | i | k | ? |
| *C. peracerosum* Gay var. *peracerosum* | + | + | + | P | i | i | k | ? |
| *C. peracerosum* var. *elegans* G. West | - | + | + | ? | ? | ? | ? | ? |
| *C. pronum* Breb. f. *pronum* | + | + | + | P | i | i | k | ? |
| *C. pronum* f. *brevius* (W. West) Kossinsk | - | + | + | ? | ? | ? | ? | ? |
| *C. rostratum* Ehr. | - | - | + | ? | ? | az | ? | ? |
| *C. setaceum* Ehr. | - | - | + | P | hb | ? | k | ? |
| *C. siliqua* W. et G. West | + | - | - | ? | ? | ? | ? | ? |
| *C. subulatum* (Kütz.) Breb. | - | + | + | ? | ? | ? | ? | ? |
| *C. tumidum* Johns. | + | + | + | ? | ? | ? | ? | ? |
| C. *venus* Kütz. f. *venus* | - | + | + | P | i | ? | k | ß |
| *C. venus* f. *minus* Roll | - | + | - | B | i | ? | k | ? |
| *Pleurotaenium ehrenbergii* (Breb.) De Bary | - | - | + | B | i | ? | k | o-x |
| *Euastrum validum* W. et G. West | + | + | + | P | hb | ? | k | ? |
| *Cosmoastrum orbiculare* (Ralfs) Pal.-Mordv. | + | + | + | P | i | ? | k | ? |
| *C. muticum* (Breb.) Pal.-Mordv. | - | + | + | B | i | ? | k | ? |
| *C. teliferum* (Ralfs) Pal.-Mordv. | - | - | + | ? | ? | ? | ? | ? |
| *Raphidiastrum longispinum* (Bail.) Pal.-Mordv. | - | - | + | ? | ? | ? | ? | ? |
| *R. lunatum* (Ralfs) Pal.-Mordv. | - | - | + | P | i | ? | b | ? |
| *Staurodesmus bilnheimeii* (Racib.) Brock | + | + | + | ? | ? | ? | ? | ? |
| *S. cuspidatus* (Breb.) Thom. | - | + | + | ? | ? | ? | ? | ? |
| *S. incus* (Breb.) Teil. var. *incus* | + | - | + | P | hb | ? | k | ? |
| *S. incus* var. *primigenius* Teil. | - | + | - | ? | ? | ? | ? | ? |
| *S. jaculiferus* (West) Teil. | - | + | - | ? | ? | ? | ? | ? |
| *S. triangularis* (Lagerh.) Teil. | - | + | + | P | i | ? | aa | ? |
| *Staurastrum anatinum* Cooke et Wille | + | - | + | ? | ? | ? | ? | ? |
| *S. arache* Ralfs | + | + | - | ? | ? | ? | ? | ? |
| *S. cyrtocerum* Breb. | - | + | - | ? | ? | ? | ? | ? |
| *S. hexacerum* (Ehr.) Wittr. | - | + | + | ? | ? | ? | ? | ? |
| *S. gracile* Ralfs | + | + | + | P | i | az | k | o-ß |
| *S. ophiura* Lund. | - | - | + | P | i | ? | k | ? |
| *S. paradoxum* Meyen | + | + | + | P | i | ? | k | ? |
| *S. sublongipes* G. M. Smith | + | + | + | ? | ? | ? | ? | ? |
| *S. subcruciatum* Cooke et Wills | - | + | - | ? | ? | ? | ? | ? |
| *S. tetracerum* Ralfs | + | + | + | ? | ? | ? | ? | ? |

| | | | | | | | | |
|---|---|---|---|---|---|---|---|---|
| *S. vestitum* Ralfs | - | - | + | P | hb | ? | k | ? |
| *Cosmarium bioculatum* Breb. | + | + | + | P | hb | ? | k | ? |
| *C. cyclicum* var. *arcticum* Nordst. f. *arcticum* | - | + | + | ? | ? | ? | ? | ? |
| *C. impressulum* Elfv. | + | + | + | B | hb | i | k | ß |
| *C. lagerheimii* Gutw. | - | + | - | ? | ? | ? | ? | ? |
| *C. punctulatum* var. *subpunctulatum* (Nordst.) Börg. | + | + | + | P | i | ? | k | ? |
| *C. subundulatum* Wille | - | + | - | ? | ? | ? | ? | ? |
| *C. subprotumidum* Nordst. | - | + | + | P | ? | az | ? | ? |
| *C. tetragonum* var. *lundellii* Cooke | - | + | + | ? | ? | ? | ? | ? |
| *C. undulatum* Corda var. *undulatum* | - | - | + | ? | i | az | k | ? |
| *C. undulatum* var. *wollei* West | - | + | + | P | i | az | k | ? |
| *Pachyphorium obsolentum* (Hantzsch) Pal.-Mordv. | + | - | - | ? | ? | ? | ? | ? |
| *Xanthidium antilopaeum* (Breb.) Kütz. | - | + | + | P | i | ? | k | o-x |
| *X. acanthophorum* Nordst. | - | - | + | ? | ? | ? | ? | ? |
| *Spondylosium planum* (Wolle) W. et G. S. West | - | + | + | P | i | ? | aa | ß |
| *S. pygmaeum* (Cooke) West | - | + | - | ? | ? | ? | ? | ? |
| *S. pulchellum* Arch. | - | + | + | ? | ? | ? | ? | ? |
| *Teilingia excavata* (Ralfs) Bourr. | + | + | + | P | i | i | k | ? |
| *T. granulata* (Roy et Biss.) Bourr. | + | + | + | P | i | i | k | ? |
| *T. wallichnii* (Jacobs.) Bourr. | - | + | - | P | i | ? | k | ? |
| *Desmidium graciliceps* (Nordst.) Lagerh. | - | + | + | ? | ? | ? | ? | ? |
| *Bambusina brebissonii* Kütz. | + | - | - | P | ? | az | ? | ? |

## 4. Conclusions

The original data helps to identify rich biodiversity of phytoplankton in the Vakh River. It includes 404 taxonomic units represented by 463 species, 463 species, types and forms of algae, belonging to 140 genera, 52 families, 13 classes and 7 divisions. In its diversity Bacillariophyta and Chlorophyta serve as the basis of phytoplankton and comprise 78.9%. As for the number of algae, Cyanobacteria, Chrysophyta and Euglenophyta play an important part (88 taxonomic units ranking lower than genus). They comprise 19.0%. The floristic importance of Xanthophyta and Dinophyta (10 species, types and forms) is relatively low with only 2.1 %.

Based on the number of taxonomic units of specific and infraspecific rank diatoms and Chlorophyta take the leading position along the river stream.

14 algae are structure-forming: vegetating throughout the whole year (*Aulacoseira italica*, *Asterionella formosa*), in the summer period (*Microcystis aeruginosa*, *Melosira varians*, *Aulacoseira granulata*, *Pandorina morum*, *Pediastrum boryanum*, *P. duplex*, *Lacunastrum gracillimum*, *Scenedesmus quadricauda*) and in the summer-fall period (*Microcystis pulverea*, *Tabellaria fenestrata*, *T. flocculosa*, *Mucidosphaerium pulchellum*).

There have been noted two new taxonomic units for subzone of middle boreal: Chrysophyta – 1, Diatoms, of genus *Eutonia* – 9, Euglenophyta – 7, Chlorophyta from *Closterium* genus – 5.

The following were identified in the Vakh River phytoplankton for the first time: 386 taxonomic units ranking lower than genus, including Chlorophyta – 159, diatoms – 135, Cyanophyta – 29, Chrysophyta- 29, Euglenophyta – 24, Xanthophyta – 7 and Dinoflagellata – 3.

The taxonomic spectrum of the Vakh River is typical for plain boreal algae plankton groups. In its composition, the dominating genera and families of phytoplankton of the Vakh River have features of undisturbed boreal northern streams. High share of families and genera with one or two species gives the River phytoplankton its northern specialty.

The taiga Vakh River gets its special features from rich genera *Closterium*, *Pinnularia*, *Desmidium*, *Eunotia* and family Desmidiaceae.

The ecological and geographical analysis demonstrated that plankton, oligogalobs and pH neutral species dominate the studied environment.

More than half of the indentified phytoplankton (51.6%) are indicative of water saprobity. The water in the Vakh River meets the requirements for β-mezosaprobian pollution zone, class of satisfactory purity (III class) based on the relationship of dominants to saprobity.

**References**

[1] Alekina O A 1953 *Fundamentals of hydrochemistry* (Leningrad: Gidrometeorol Publishing House)
[2] Barinova S S, Medvedeva L S and Anisimova O V 2006 *Biodiversity of algae indicators* (Tel Aviv: Pilies Studio)
[3] Beirom S G 1975 *Resources of surface water in USSR* **15 (1)** 408
[4] Vasiljeva I I 1989 *Analysis of species and algae development in water bodies in Yakutiya* (Yakusk)
[5] Vasser S P 1989 *Algae: Reference book* (Kiev: Naukova Dumka)
[6] Vorobjova S S 1995 *Phytoplankton of Angara river water bodies* Abstract of Thesis for the Degree of the Candidate of Biological Sciences (Novosibirsk)
[7] Gabyshev V A 1999 *Plankton algae in the Lena River in the area influenced by Yakutsk ecosystem* Abstract of Thesis for the Degree of the Candidate of Biological Sciences (Yakutsk)
[8] Genkal S I, Bondarenko N A and Schur L A 2011 *Bacillariophyta of lakes of the South and the North of Eastern Siberia* (Rybinsk: Rybinsk Publishing House)
[9] Getsen M V 1985 *Algae in the ecosystems of the Far North* (Leningrad: Nauka Publishing house)
[10] Gollerbakh M M and Polyanski V I 1951 *Field guide of freshwater algae (General information)* **1** (Moscow: Sovetskaya Nauka)
[11] Makarova I V 1988 *Diatoms of the USSR (fossil and living)* **2 (1)** (Leningrad: Nauka)
[12] Zabelina M M, Kiselev I A et al 1951 *Field guide of diatoms Diatoms – Bacillariophyta* **4** (Moscow: Sovetskaya Nauka)
[13] Kondratyeva N V 1972 *Morphologyand symantics of Hormogoniophyceae algae that leads to algal bloom of Dniepr and its reservoirs* (Kiev: Naukova Dumka)
[14] Kondratyeva N V 1975 *Morphogenesis and main evolution means of Hormogoniophyceae algae* (Kiev: Naukova Dumka Publishing House)
[15] Kondratyeva N V 1981 *Botanicheskij zhurnal* **66 (2)** 215–26
[16] Kondratyeva N V, Kovalenko O V and Prokhodjkova L P 1984 *Cyanobacteria Cyanophyta. Part 1. Field guide of fresh water algae of Ukrainian SSR* (Kiev: Naukova Dumka)
[17] Kopyrina L I 2009 *Proceedings of the 2nd national conference "Algae: issues of taxonomy, ecology and monitoring" (Syktyvkar: 5-9 October, 2009)* (Syktyvkar) p 197–8
[18] Vingerg G G 1981 Guidelines on collection and processing of samples in hydrobiological studies in fresh water (Leningrad: State Research Institute of River and Pond Fishery)
[19] Naumenko J V 1985 *Phytoplankton of Ob' and Lower Irtysh and its changes under the influence of anthropogenic factors* Abstract of Thesis for the Degree of the Candidate of Biological Sciences (Novosibirsk)
[20] Naumenko J V 1991 *Siberian Ecological Journal* **2** 28–33
[21] Naumenko J V 2001 *Botanical studies of Siberia and Khazakhstan* **7** 43–49
[22] Naumenko J V 1996 *Phytoplankton of the Ob' River* Abstract of Thesis for the Degree of the Doctor of Biological Sciences (Novosibirsk)
[23] Naumenko J V 1998 *Siberian Ecological Journal* 2 147–150
[24] Naumenko J V and Skorobogatova O N 2009 *Turczaninowia* **12 (1-2)** 65–70
[25] Naumenko J V and Skorobogatova O N 2009 *Bulletin of Ecology, Forest Science and Landscape Ecology* **10** 49–53
[26] Palamar-Mordvintseva G M 1982 *Field guide fresh water Algae of USSR. Issue 11(2). Chlorophyta. Conjugatophyceae Class. Desmidiáles Order* (Leningrad: Sovetskaya nauka)

[27] Remigaylo P A *1995 Phytoplankton of the River Viluj and Viluj Reservoir* Abstract of Thesis for the Degree of the Candidate of Biological Sciences (Novosibirsk)
[28] Ivanov Yu N 1974 *Resources of surface water in USSR. General hydrological characteristics (1963 – 1970)* **15** **(1)** (Leningrad)
[29] Safonova T A and Shaulo S P 2006 *Turczaninowia* **9** **(3)** 102–8
[30] Skorobogatova O N 2010 *Natural and Technical Sciences* **3** 107–11
[31] Fjodorov V D 1979 *On study methods of phytoplankton and its activity* (Moscow: MSU Publishing House)
[32] Tsarenko P M 1990 *Brief field guide of Chlorococcales of Ukrainian SSR* (Kiev: Naukova Dumka)
[33] Tsarenko P M 2005 *Algology* **15** **(4)** 459–67
[34] Elyashev A A 1957 *Collection of research papers on palaeontology and biostratigraphy* **4** 123–37
[35] Bukhtiyarova L and Round F E 1996 *Diatom Research* **11** 1–30
[36] Hegewald E, Hanagata N and Tsarenko P 2000 *Algologia* **10** **(4)** 176
[37] Kharitonov V G and Genkal S I 2012 *Diatoms of the Elgygytgyn Lake and ist Vicinities (Chukotka)* (Magadan: NESC FEB RAS)
[38] Lange-Bertalot H and Krammer K 2002 *Diatoms of Europe: Diatoms of the European Inland Waters and Comparable Habitats. Cymbella (Vol. 3)* Gantner Publishing
[39] Lange-Bertalot H and Krammer K 2000 *Diatoms of Europe: Diatoms of the European Inland*
[40] *Waters and Comparable Habitats. The genus Pinnulara (Vol. 1)* Gantner Publishing
[41] Lange-Bertalot H and Krammer K 2001 *Diatoms of Europe: Diatoms of the European Inland*
[42] *Waters and Comparable Habitats. Navicula sensu stricto, 10 genera separadet from Navicula sensu strict. Frustulia* Gantner Publishing
[43] Dorofeyuk N I and Kulikovskiy M S 2012 *Diatoms of Mongolia* (Moscow)

IOP Conf. Series: Earth and Environmental Science **138** (2018) 012018 doi:10.1088/1755-1315/138/1/012018

# Bryoflora of mountain mires of Sub-Polar Urals

**Yu V Skuchas**[1,3], **E D Lapshina**[2,4]

[1]Museum of Nature and Man, Mira str. 11, Khanty-Mansiysk, 628011, Russian Federation
[2]Yugra State University, Chekhova str., 16, Khanty-Mansiysk, 628012, Russian Federation

[3]jskuchas@umuseum.ru
[4]e_lapshina@ugrasu.ru

**Abstract**. The Bryoflora of mountain mires of Sub-Polar Urals within the Khanty-Mansi Autonomous District was studied in 2013-2015. The field work was conducted at Mt. Ner-Oika and in the upper course of the Puiva River (64° 27' - 64° 35' N, 59°35' - 59° 46'E). In total, 76 species of mosses were identified in the 24 mires surveyed. Most mosses have specific habitat preferences to different mire types, 5 species were observed almost in all types of mires: *Warnstorfia exannulata, Straminergon stramineum, Sanionia uncinata, Pohlia nutans*, and *Pleurozium schreberi*. A significant number of rare moss species were identified among the mires bryoflora (*Dicranum laevidens, D. leioneuron, Loeskypnum badium Paludella squarrosa*, *Philonotis tomentella* and others). Two species, *Tayloria lingulata* and *Philonotis serotina* were observed in the Khanty-Mansi Autonomous District for the first time.

## 1. Introduction

Sub-Polar Urals (SPU) is the highest and most wide-stretching part of the Ural Mountains located between 65°40' and 64° N. SPU is the location of Urals' highest peaks (incl. Mt. Narodnaya, 1894 m a.s.l.). In contrast with the rest of the Urals, SPU commonly features Alpine-type landscapes characterized by well-defined high mountain peaks, sharp ridges, steep slopes, and deep valleys. The axial part of the mountains in this area is adjoined by 360 – 600 m high foothills alternating with lower flats. The Urals stretch from north to south blocking movement of relatively warm humid Atlantic air masses and making their eastern slopes of SPU much drier and colder than the western ones. Precipitation does not exceed 800 mm per year in the uplands whereas it can be 1500 mm on the western slopes. The eastern slopes are in the area of influence of the Siberian anticyclone with clear sunny weather being access to the masses of Arctic air from the Barents and Kara seas. It leads to a mixture of Arctic and continental influences and a cold climate [1]. Climatic features of the area can be summarized as follows: the climate is continental with long winters and very short summers. The mean temperature of January is –19-23°C, minimal values reach –50-55°C. The mean temperature of July varies from 14-16°C in plains to 5° at height of 1300-1400 m. There are large permanent snowfields on the leeward eastern slopes of the highest peaks [2, 3].

Due to the difficulty of access, the SPU area is dominated by pristine natural landscapes, the biodiversity of which has been poorly characterized in terms of both species and biocenoses, including bryophytes. The moss flora of SPU has been investigated during more than 150 years. The previous reseach of this region was mainly conducted on the forested mountain slopes and on neighboring lowlands, both on the west and east slopes [4, 5]. The data on moss flora from numerous publications [6-13] was summarized at the end of last century [14]. The aim of this paper is to present the results of our study on the moss species composition of mountain mires of SPU as a very specific type of habitat with a high concentration of species extremely rare in the plains of Western Siberia.

## 2. Material and methods

Here we report on the field work conducted in 2013-2015 at Mt. Ner-Oika and in the upper course of the Puiva River (64° 27' - 64° 35' N, 59°35' - 59° 46'E). Both sites are located in proximity to the drainage divide of the Urals on its eastern slope 380 to 1200 m a.s.l (figure 1). The research area covers the upper part of the forest zone, the mountain-forest-tundra zone, and a mountain tundra zone transitioning to rocky deserts dominated by loose rock screes (kurumniks). The upper limit of dark coniferous forest is situated at 400-500 m a.s.l. The mountain-forest-tundra zone is represented by larch-dominated open woodlands, secondary low birch forests, alder tickets and subalpine meadows. Since 500-600 m, mountain tundra forms the belt: draft shrub - green moss and moss lichen tundra in the lower part is replaced by dry debris-lichen-moss tundra, which is succeeded by rock fields with only scattered and fragmentary vegetation at the level above 900-1000 m a.s.l. However, due to the highly dissected relief, the location of the upper limit of the subalpine wood communities is quite variable in the SPU [3].

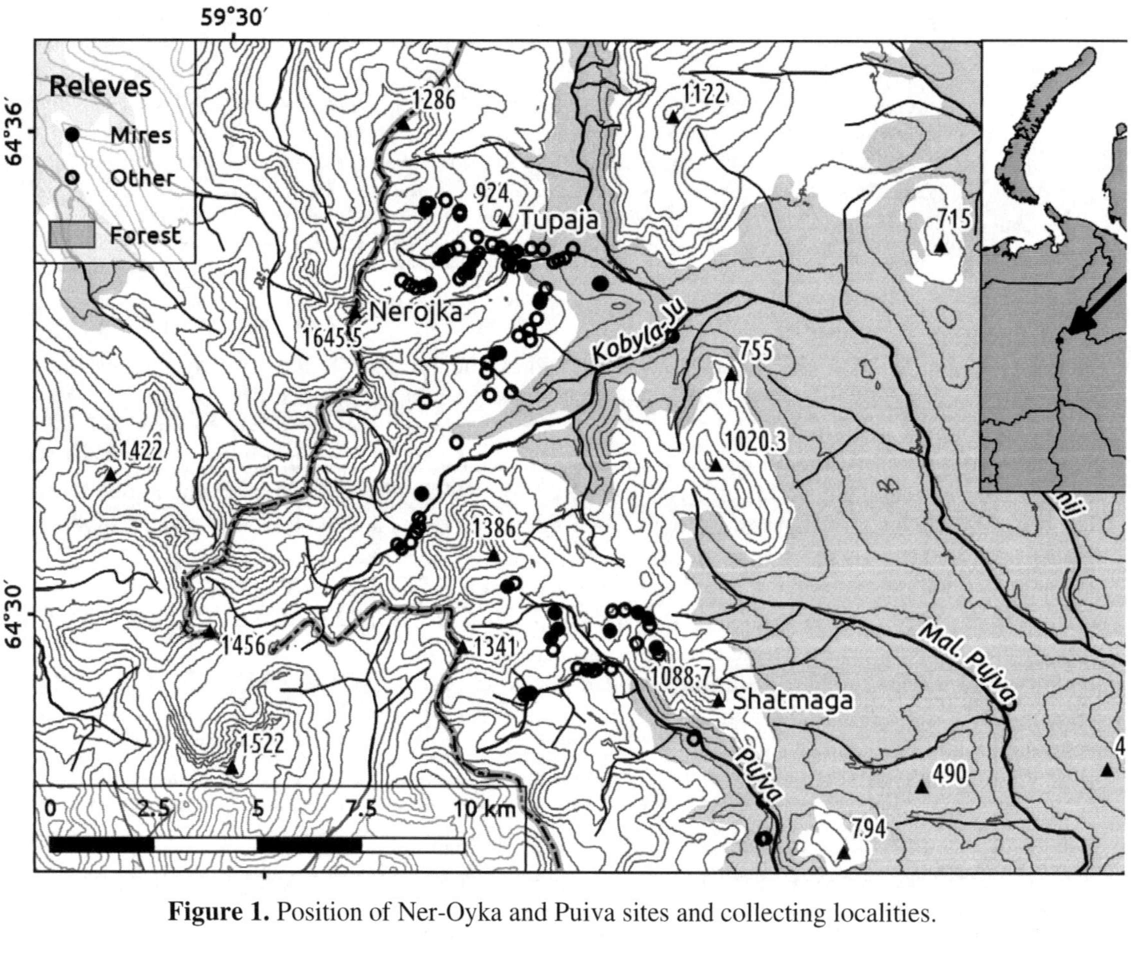


**Figure 1.** Position of Ner-Oyka and Puiva sites and collecting localities.

In two chosen key sites, 24 relevés were done, attempting to cover different types of mires and to reveal species diversity. The relevés were accompanied by collecting moss samples. In total, 242 specimens were collected: 127 at Mt. Ner-Oika and 115 at the upper course of the Puiva River.

A preliminary list of species and their abundance have been drafted in field. The identifications of the speciments were checked later in the laboratory. All herbarium samples collected during the field work are stored in the herbarium of the Yugra State University (YSU) and duplicates are in the Museum of Nature and Man in Khanty-Mansiysk.

## 3. Results and Discussion

The mires in the study area are rather small (from several square meters to one-two hectares) and totally occupy not more than 1% of the land area [4, 5]. Nevertheless, a significant species diversity is present due to the favorable combination of water and mineral nutrition, humidity conditions, the presence of peat or peat-mineral soils. Three types of mires were distinguished: mountain dwarf shrubs-sedge-sphagnum bogs predominantly with rain water-supply, rich fens with groundwater water-supply, and transitional sedge-sphagnum mires. 76 species of mosses were identified in the 24 mires surveyed (30% of the total bryoflora in the study area).

Mountain dwarf shrubs-sedge-sphagnum bogs feature high diversity of moss species (41 species, representing 54% of the total moss species identified for all types of mires). Sphagnum mosses predominate and cover up to 95% of the surface, including various combinations in different plant communities of *Sphagnum russowii, S. angustifolium, S. capillifolium, S. balticum, S. girgensohnii, S. warnstorfii,* and *S. lindbergii. Sphagnum fuscum, S. teres,* and *S. aongstroemii* are less common. Green mosses are well-represented (up to 20%), most commonly *Polytrichum strictum, Aulacomnium turgidum, Dicranum elongatum*, and *D. laevidens* (table 1).

Transitional sedge-sphagnum mires of the upper part of the forest zone lack a distinctive species diversity (20 species or 26%). Their species composition is similar to the bogs of the first group (table 1), however several notable differences exist. Sphagnum mosses cover almost all the surface, dominated by *Sphagnum riparium* and *S. fallax* with either minor or co-dominant presence of *Sphagnum jensenii, S. flexuous*, and *S. fimbriatum.*

Rich fens with groundwater water-supply feature the similar diversity of species (42 species, representing 55% of the total moss species identified). Sphagnum mosses are poorly represented, the only constantly encountered species being *Sphagnum warnstorfii,* occasionally accompanied by small numbers of *S. squarrosum* and *S. girgensohnii*. Rich fens biodiversity is characterized by such species as *Calliergon richardsonii, Bryum pseudotriquetrum, Paludella squarrosa, Aulacomnium palustre, Tomentypnum nitens, Helodium blandowii, Campylium stellatum, Plagiomnium ellipticum,* and *Rhizomnium pseudopunctatum* (table 1). Such species as *Tayloria lingulata* and *Philonotis serotina* were exclusive to this type of mires.

The following 5 species were observed in all types of the mires: *Warnstorfia exannulata, Straminergon stramineum, Sanionia uncinata, Pohlia nutans,* and *Pleurozium schreberi* (table 1).

**Table 1.** Frequency of moss species on studied mountain mires of Sub-Polar Urals. Mire types: 1 – Mountain dwarf shrubs-sedge-sphagnum bogs, 2 – Transitional sedge-sphagnum mires, 3 – Rich fens. Frequency: + – single, I – 1-20%, II – 1-40%, III – 41-60%, IV – > 60%.

| Moss species | 1 | 2 | 3 |
|---|---|---|---|
| *Sphagnum russowii* | IV | III | |
| *Sphagnum angustifolium* | III | IV | |
| *Sphagnum girgensohnii* | III | IV | I |
| *Polytrichum strictum* | IV | III | |
| *Polytrichum commune* | III | III | |
| *Sphagnum lindbergii* | III | II | |

| | | | |
|---|---|---|---|
| *Sphagnum teres* | II | II | |
| *Sphagnum balticum* | III | | |
| *Aulacomnium turgidum* | III | | |
| *Dicranum laevidens* | III | | |
| *Dicranum elongatum* | III | | |
| *Dicranum leioneuron* | II | | |
| *Sphagnum capillifolium* | III | | |
| *Sphagnum fuscum* | II | | |
| *Polytrichastrum alpinum* | II | | |
| *Sphagnum aongstroemii* | II | | |
| *Loeskypnum badium* | II | | |
| *Polytrichum jensenii* | I | | |
| *Sphagnum compactum* | I | | |
| *Sphagnum contortum* | + | | |
| *Sphagnum majus* | + | | |
| *Sphagnum riparium* | | IV | |
| *Sphagnum fallax* | | III | |
| *Sphagnum jensenii* | | III | |
| *Sphagnum flexuosum* | | II | |
| *Sphagnum fimbriatum* | | II | |
| *Polytrichastrum longisetum* | | II | |
| *Sphagnum warnstorfii* | II | | IV |
| *Aulacomnium palustre* | | | V |
| *Bryum pseudotriquetrum* | | | III |
| *Calliergon richardsonii* | | | III |
| *Rhizomnium pseudopunctatum* | | | III |
| *Tomentypnum nitens* | | | III |
| *Hylocomiastrum pyrenaicum* | | | III |
| *Philonotis tomentella* | | | III |
| *Philonotis fontana* | | | II |
| *Sciuro-hypnum latifolium* | | | II |
| *Paludella squarrosa* | | | II |
| *Plagiomnium ellipticum* | | | II |
| *Bryum weigelii* | | | II |
| *Helodium blandowii* | | | II |
| *Campilium stellatum* | | | II |
| *Warnstorfia exannulata* | II | II | III |
| *Straminergon stramineum* | III | II | IV |
| *Pohlia nutans* | III | II | II |
| *Sanionia uncinata* | II | I | III |
| *Pleurozium schreberi* | II | II | I |
| *Warnstorfia sarmentosa* | II | | II |
| *Dicranum bonjeanii* | I | | II |
| *Oncophorus wahlenbergii* | II | | I |
| *Scorpidium revolvens* | I | | I |
| *Hylocomium splendens* | I | | I |

In addition, the following species were observed: 1 (mountain dwarf shrubs-sedge-sphagnum bogs) - *Conostomum tetragonum I, Dicranum angustum +, D. flexicaule +, D. majus +, D. spadiceum +,*

*Kiaeria blyttii* +, *Racomitrium lanuginosum* +, *Warnstorfia fluitans* +; 2 (rich fens) - *Calliergonella lindbergii* +; 3 (transitional sedge-sphagnum mires) - *Brachythecium mildeanum I, B. rivulare I, Campylidium sommerfeltii I, Cinclidium dendroides I, Dicranella schreberiana I, Dicranum fuscescens* +, *Mnium stellare I, Philonotis seratina I, Plagiothecium denticulatum I, Pohlia cruda I, P. wahlenbergii I, Pseudobryum cinclidioides I, Rhizomnium magnifolium I, Sciuro-hypnum starkei I, Tayloria lingulata I.*

While the transitional sedge-sphagnum mires are located at the upper part of the forest zone, two other types of mires are not rare along a whole slope up from forest zone to the upper limit of tundra (at 915-960 m a.s.l.). In addition to the typical bog species, *Polytrichum commune, Polytrichastrum alpinum* are infrequently encountered in mountain dwarf shrubs-sedge-sphagnum bogs. *Sphagnum contortum, Loeskypnum badium, Dicranum leioneuron,* and *Polytrichum jensenii* were exclusive to this type of mires. Rich fens' bryoflora is additionally enriched by mountainous species such as *Hylocomiastrum pyrenaicum,* and *Sciuro-hypnum latifolium* (at the level above 700 m a.s.l.), as well as by the presence of *Philonotis fontana, P. tomentella,* and *Bryum weigelii* which indicates free-flowing channel runoff in the area of their growth.

A significant number of rare moss species were identified among the mires bryoflora. Some of them, *Calliergon richardsonii, Paludella squarrosa, Dicranum laevidens, D. leioneuron,* and *Loeskypnum badium,* are quite common in SPU bogs and have a significant function in phytocenoses formation there. Two species, *Tayloria lingulata* and *Philonotis serotina* were observed in the Khanty-Mansi Autonomous District for the first time and deserve to be added to the regional Red List of Threatened species.

**Acknowledgements**
Authors are sincerely grateful to Dr. M. Ignatov and Dr. E. Ignatova for their help in checking identifications of some specimens.

**References**
[1] Kuvaev V B 2006 *The flora of subarctic mountains in Eurasia and altitudinal distribution of its species* (Moscow: KMK Scientific Press Ltd.) p 568
[2] Kemmerikh A O 1970 *Sub-Polar Urals* (Moscow: Fizkultura I sport) p 155
[3] Konstantinova N A and Lapshina E D 2014 On the hepatic flora of the eastern Subpolar Ural (Khanty-Mansi autonomous district) *Arctoa* **23** 80–90
[4] Lapshina E D, Ignatova E A, Ignatov M S and Skuchas Yu V 2015 Contributions to the knowledge of moss flora of the Subpolar Ural mountains highland (Khanty-Mansi Autonomous District) *Proc. of the Int. Bryological Conf. devoted to the 100-th Anniversary of Anastasiya Lavrentievna Abramova (Saint Petersburg, October 12-16, 2015)* ed I V Czernyadjeva (St. Petersburg: Publishing house of ETU) pp 96–100
[5] Lapshina E D, Noskova M G and Skuchas Yu V 2016 Sphagnum mosses in vegetation cover of highlands in the Sub-Polar Ural, Khanty-Mansiysk Autonomous District *Arctoa* **25** 177–182
[6] Pohle R R 1915 Contributions to the vegetation of northern Russia. I. On the moss flora on northern Russia *Trudy Iperatorskogo Botanicheskogo Sada* **33** I-VIII 1–148
[7] Ivanovsky V 1917 List of the mosses collected by B N Gorodkov at Severnaja Sosva river's basin (Berezovsky district, Tobolsk province) *Ezhegodnik Tobolskogo gubernskogo muzeja, god 26* **28 (2)** 1–12
[8] Ghaze O F 1930 The list of peat mosses collected by B N Gorodkov in the Western Siberia and by B N Gorodkov and V B Soczava in the Polar Urals *Trudy Botanicheskogo Museya Akademii Nauk USSR* **22** 207–213
[9] Soczava V B 1930 The forests' limits in Ljapin Ural mountains *Trudy Botanicheskogo Museya Akademii Nauk USSR* **22** 1–48
[10] Zinzerling Yu D 1935 An overview of vegetation of Sablja Mountain Range *Trudy lednikovykh expeditsij* **4** *Ural Pripolyarnyje raiony* 75–86

[11] Gorchakovsky P L 1958 Vegetation of Sablja range at the Subarctic Urals *Rastitelnost krainego severa USSR i ee osvoenije* **3** 95–127
[12] Kuvaev V B 1970 Lichens and mosses on the Sub-Polar Urals and adjoined plains *Trudy Instituta Ecologii Rastenij I Zhivotnykh Uralskiy filial Akademii Nauk USSR* **70** 61–92
[13] Gorchakovsky P L 1975 *Vegetation of the high-mountain Urals' regions* (Moscow: Nauka) p 283
[14] Dyachenko A R 1995 History of the studies of moss flora in the Subarctic Urals *Arctoa* **4** 17–22